Universität Stuttgart
Institut für Energieübertragung und Hochspannungstechnik, Band 45

Thermal Modelling of Power Transformers
Using Computational Fluid Dynamics

FSC
www.fsc.org
MIX
Papier aus ver-
antwortungsvollen
Quellen
Paper from
responsible sources
FSC® C105338

Thermal Modelling of Power Transformers Using Computational Fluid Dynamics

Von der Fakultät

Energie-, Verfahrens- und Biotechnik

der Universität Stuttgart

zur Erlangung der Würde eines Doktor-Ingenieurs (Dr.-Ing.)

genehmigte Abhandlung

vorgelegt von

Saeed Khandan Siar

aus Teheran, Iran

Hauptberichter:　　　　　Prof. Dr.-Ing. Stefan Tenbohlen

Mitberichter:　　　　　Prof. Dr.-Ing. Stefan Riedelbauch

Tag der mündlichen Prüfung:　10.04.2024

Institut für Energieübertragung und Hochspannungstechnik

der Universität Stuttgart

2024

Bibliografische Information der Deutschen Nationalbibliothek:

Die Deutsche Nationalbibliothek verzeichnet diese Publikation in der Deutschen Nationalbibliografie, detaillierte bibliografische Daten sind im Internet über http://dnb.dnb.de abrufbar.

Universität Stuttgart

Institut für Energieübertragung und Hochspannungstechnik, Band 45

D 93 (Dissertation Universität Stuttgart)

Thermal Modelling of Power Transformers Using Computational Fluid Dynamics

Autor: Saeed Khandan Siar

Verlag: BoD • Books on Demand GmbH, In de Tarpen 42, 22848 Norderstedt

Druck: Libri Plureos GmbH, Friedensallee 273, 22763 Hamburg

ISBN: 978-3-7578-5192-7

Acknowledgement

This study was carried on during my work as an academic employee at the Institute of Power Transmission and High Voltage Technology (IEH) at the University of Stuttgart in Germany. While studying and researching at the university, I have experienced great assistance and enjoyable moments with my colleagues.

I would like to thank my supervisor, the head of the institute Prof. Dr. -Ing. Stefan Tenbohlen for giving me a unique opportunity to enhance my knowledge and research with his academic guidance and personal encouragement within the last six years. Besides, I express my deepest gratitude to Dr. Ulrich Schärli for his continuous support and encouragement throughout my academic journey at the institute. Moreover, I want to thank Mrs. Schärli, Mrs. Janzen, for their persistent support in enhancing my proficiency in the German language.

Regarding my research, I would like to thank the technical staff of the institute and workshop colleagues. With the comprehensive experiences of these experts, I could perform my experimental tasks easier. Certainly, I would like to thank Dr. Banka and Mr. Beura for our fruitful discussions and continuous collaborations. I feel fortunate to have the honour to be a member of an outstanding group that was always ready to assist and share their knowledge and experiences. Their support is gratefully acknowledged.

Most of all, I want to express my sincere appreciation to my lovely wife, Banafsheh. With her outstanding support, I could continue to reach my goals, and with her great patience, I gradually learned how to face and solve new tasks in my life. I have made it with her faithful assistance. Furthermore, I am deeply grateful to my family; my parents, Nahid and Hamid, my kind sister Sahar and last but certainly not least, Daniel, my great supportive friend likewise my family member. I consider myself very lucky to have these people in my life.

Abstract

Power grids have improved rapidly due to the latest developments in energy transmission and the rapid expansion of renewable energy uses. Power transformers have become vital equipment in providing a sustainable power network in energy transmission and distribution systems. In this regard, it is essential to minimize the harmful effects of thermal stresses in power transformers during the lifetime of the power transformer. Therefore, identifying the source of heat losses within the transformer is the primary step, followed by determining the optimized cooling systems that can increase the lifetime of the power transmission systems and allowable loading capacity. This is because thermal stresses in the windings of a power transformer can increase the failure rate during operations, leading to accelerated thermal ageing and shortened lifetime. As higher temperatures within the power transformers can result in faster material degradation, it is crucial to address the thermal stresses to ensure the reliability of the power transformer.

This study examines the thermal behavior of power transformers using the Computational Fluid Dynamics (CFD) numerical method. First, the investigation is performed based on measured data from real power transformers and winding models in the laboratory and the corresponding numerical CFD winding models suitable for the investigations are created and a mesh sensitivity analysis is carried out to ensure their accuracy before using the numerical models. Furthermore, the effects of changing oil temperature on the thermal driving force are considered since oil materials have temperature dependent characteristics. Overall, this research aims to gain insights into the thermal behavior of power transformers to improve their lifetime and reliability in energy transmission and distribution systems.

Furthermore, the state-of-the-art of transformers is reviewed first, discussing the motivation behind the study and introducing the basic principles of power transformers and the main concepts of cooling methods. Then, the governing equations and the fundamental principles of fluid flow and heat transfer are explained. The sources of power losses within transformers are introduced as initial boundary conditions during investigations. Furthermore, this work highlighted the impact of generated heat on the thermal behavior of the transformers.

Since 3D CFD calculations display an excellent agreement with the measurement data under different operational conditions, they allow the determination of the temperature and fluid flow distribution within winding models. This work considers popular cooling methods including natural and force convectional fluid flow within oil domain.

It examines operational parameters like fluid flow rate and inlet temperature to determine the significant influence of the Reynolds and Prandtl numbers. Based on the result, we can conclude that the Reynolds number dominates in forced cooling mode. In the meantime, this study employs a geometrical study to identify the effects of different dimensions on the vertical and horizontal cooling channels to determine the reliability and accuracy of the numerical CFD calculations. In this light, this study establishes that the lower height of the horizontal channels leads to a better cooling performance of the winding model. Furthermore, reducing eddies at the inlet of the horizontal channels leads to a more uniform oil distribution within the cooling channels.

This study also examines the thermal performance of using biodegradable oil material in OD cooling mode. Along with the geometrical design of the windings, the winding temperature and oil flow distributions of different oil materials are investigated. It is determined that natural ester oil has a lower average winding temperature than mineral oil under force cooling conditions with an OD structure. Moreover, the study observes that the oil distribution of natural ester through the cooling channels is more evenly under identical operational condition. It also provided a more uniform share of oil for horizontal cooling channels. In this light, while the position of the hot spot temperature remains unchanged, the use of natural ester oil results in a significant reduction of hot spot temperature in OD cooling mode.

This study analyses the effects of non-uniform heat loss distributions, which are very common in the operating conditions of power transformers. Higher heat losses lead to higher average and hot spot temperatures at different oil inlet temperatures of the windings. For a uniform heat loss distribution, the location of the hot spot temperature depends directly on oil flow distribution. However, considering the non-uniform heat loss distribution, the location of hot spot temperature might change. The study also examines the impact of a winding design equipped with additional vertical cooling channel, including an extra vertical cooling channel in the middle of the pass.

The results show that using additional vertical cooling channel reduces the average winding temperature and keeps the hot spot temperature lower, effectively reducing the degradation in disc type windings. In addition to conducting steady-state analyses, this study analyses the transient thermal behavior of power transformers in natural cooling mode. The research also considers the buoyancy effect over time and calculates hot spot factors for various winding designs. During the transient thermal behaviour, it is observed that the rate of the temperature rise at the initial time is more pronounced. As the temperature increases, the viscosity of the oil decreases and the cooling oil flows throughout all the channels.

Kurzfassung

Aufgrund der zunehmenden Anzahl dezentraler Energieerzeugungsanlagen, die auf erneuerbaren Energiequellen basieren, ist ein Ausbau der Stromnetze erforderlich. Leistungstransformatoren sind im Stromnetz Schlüsselkomponenten und haben einen maßgeblichen Einfluss auf die Zuverlässigkeit der Übertragungs- und Verteilnetze. In diesem Zusammenhang ist es von entscheidender Bedeutung, die schädlichen Auswirkung von thermischen Belastungen in Leistungstransformatoren während ihrer Lebensdauer zu minimieren. Zur Erhöhung der Lebensdauer und der zulässigen Belastbarkeit von Leistungstransformatoren gilt es die Kühlsysteme zu optimieren.

Thermische Belastungen in den Wicklungen sowie höhere Temperaturen innerhalb des Transformators führen zu einer beschleunigten thermischen Alterung und begünstigen daher eine schnellere Materialalterung. Zur Sicherstellung der notwendigen Zuverlässigkeit eines Leistungstransformators ist die Minimierung der thermischen Belastung in den Windungen entscheidend. Darüber hinaus wird die Motivation der wissenschaftlichen Ausarbeitung dargelegt und eine Einführung in die Grundprinzipien der Leistungstransformatoren mit den wichtigsten Konzepten der Kühlmethoden gegeben.

In dieser Dissertation wird das thermische Verhalten von Leistungstransformatoren mittels numerische Strömungssimulation (CFD) behandelt. Die Untersuchung wird zunächst basierend auf Messdaten realer Leistungstransformatoren und Wicklungsmodellen aus dem Labor durchgeführt. Anschließend werden entsprechende numerische CFD-Wicklungsmodelle erstellt und eine Netzempfindlichkeitsanalyse durchgeführt. Dies soll die Genauigkeit der numerischen Berechnungen der Simulationen sicherstellen.

Darüber hinaus werden die Auswirkungen der Variation der Öltemperatur auf die thermische Antriebskraft und die Dichteänderung des Öls berücksichtigt, da Ölmaterialien temperaturabhängige Eigenschaften aufweisen. Insgesamt zielt diese Forschung darauf ab, Einblicke in das thermische Verhalten von Leistungstransformatoren zu gewinnen, um ihre Lebensdauer und Zuverlässigkeit in Energieübertragungs- und Verteilungssystemen zu optimieren.

Diese Arbeit bietet einen Überblick über den aktuellen Stand der Technik in der thermischen Modellierung von Leistungstransformatoren. Anschließend werden die grundlegenden Gleichungen und Prinzipien der Fluidströmung und der Wärmeübertragung erläutert. Die Quellen der Leistungsverluste in Transformatoren werden während den numerischen Untersuchungen als Anfangsrandbedingungen der mehrdimensionalen Differentialgleichungen eingesetzt. Außerdem wird der Einfluss der erzeugten Wärme auf das thermische Verhalten der Transformatoren dargestellt.

Da die dreidimensionalen CFD-Berechnungen in dieser Arbeit eine hervorragende Übereinstimmung mit den Messdaten aus den verschiedenen Betriebszuständen aufweisen, ist die Bestimmung der Temperaturverteilung und der Strömungsverteilung innerhalb der Wicklungsmodelle möglich.

Im Rahmen dieser Promotionsarbeit werden die bekanntesten Kühlmethoden, einschließlich der natürlichen und erzwungenen Konvektionsströmung, berücksichtigt. Als Betriebsparameter werden beispielsweise die Durchflussmenge und die Öleintrittstemperatur untersucht, um den Einfluss der Reynolds- und Prandtl-Zahlen zu ermitteln. Dabei wurde festgestellt, dass die Reynolds-Zahl eine maßgebliche Rolle im Konzept des erzwungenen Kühlmodus einnimmt. Gleichzeitig wurden zur Bestimmung der Zuverlässigkeit und der Genauigkeit der CFD-Berechnungen geometrische Studien durchgeführt. Diese geometrischen Studien ermöglichen die Identifikation der Auswirkungen der unterschiedlichen Abmessungen der vertikalen und horizontalen Kühlkanäle.

In diesem Zusammenhang ergibt sich aus den Untersuchungen die Erkenntnis, dass eine geringere Höhe der horizontalen Kanäle zu einer besseren Kühlleistung des Wicklungsmodells führt. Darüber hinaus wird dieses Design nicht nur die Strömungswirbel beim Eintritt in die horizontalen Kanäle reduziert, sondern ermöglicht auch eine gleichmäßigere Ölverteilung innerhalb der Kühlkanäle.

Diese Studie untersucht ebenfalls die thermische Leistungsfähigkeit von biologisch abbaubarem Ölmaterial im OD-Kühlmodus in Verbindung mit dem geometrischen Design der Wicklungen. Die Ölflussverteilungen sind bei Anwendung verschiedener Ölmaterialien unterschiedlich. Natürliches Esteröl erreicht eine ähnliche durchschnittliche Wicklungstemperatur wie Mineralöl, vorausgesetzt, dass das Konzept der forcierten Kühlungssysteme mit ölgeleiteter Struktur angewendet wird.

Darüber hinaus ist bei gleichen Betriebsbedingungen die Ölverteilung des natürlichen Esteröls durch die Kühlkanäle homogener.

Dies sorgt für eine gleichmäßigere Ölverteilung in den horizontalen Kühlkanälen. Die Position der Heißpunkttemperatur wird zwar nicht signifikant beeinflusst, jedoch führt die Anwendung von natürlichem Esteröl zu einer Senkung der Heißpunkttemperatur.

Die Einflüsse ungleichmäßiger Wärmeverlustverteilungen, die unter den Betriebsbedingungen von Leistungstransformatoren sehr häufig vorkommen, werden ebenfalls in dieser Arbeit untersucht. Höhere Wärmeverluste führen zu höherem Durchschnitt und Heißpunkttemperaturen bei verschiedenen Öleinlauftemperaturen in den Wicklungen. Bei einer gleichmäßigen Wärmeverlustverteilung hängt die Position der Heißpunkttemperatur direkt von der Ölflussverteilung ab. Bei einer ungleichmäßigen Wärmeverlustverteilung kann sich die Position der Heißpunkttemperatur jedoch ändern.

Darüber hinaus wird das Wicklungsdesign mit einem Noppenband (als zusätzlicher vertikaler Kanal) untersucht. Der zusätzliche vertikale Kühlkanal im mittleren Bereich der Passage senkt die durchschnittliche Wicklungstemperatur und hält die Heißpunkttemperatur bei gleichen Betriebsbedienung auf einem niedrigeren Niveau.

Daher wird durch die Verwendung eines Noppenbands die Degradierung der thermischen Leistung von Scheibenwicklungen wirksam reduziert. Zusätzlich zur stationären Analyse des thermischen Verhaltens von Leistungstransformatoren wird das instationäre thermische Verhalten im natürlichen Abkühlungsmodus numerisch und experimentell geprüft. Die Auswirkung der thermischen Antriebskräfte, der sogenannte Auftriebseffekt, wird transient simuliert. Die Berechnung der Heißpunktfaktoren wird für verschiedene Wicklungskonstruktionen dargestellt.

Die Auswertung zeigt, dass die Heißpunktfaktoren im ersten Zeitintervall der Untersuchung zunehmen und dann aufgrund der höheren Ölgeschwindigkeiten in den Kanälen abnehmen. Außerdem ist der Gradient der steigenden Heißpunkttemperatur im ersten Zeitintervall höher. Sobald die Ölviskosität mit steigender Temperatur abnimmt, fließt das Öl schneller durch alle Kanäle.

Table of Contents

List of Figures

List of Tables

Nomenclature

Latin symbols

Symbol	Dimension	Description
A	m^2	area
a	m^2/s	thermal diffusivity
A_ϕ	m^2	cross section
$A_{channel}$	m^2	cross section of vertical cooling channel at inner winding diameter
a_{rad}	–	absorptivity
c	$J/kg \cdot K$	specific heat capacity
cp	$J/kg \cdot K$	isobaric specific heat capacity
$\vec{D}$	N/m^2	deformation gradient tensor
d	m	diameter
D_{hyd}	m	hydraulic diameter
d_{in}	mm	inner winding diameter
d_{out}	mm	outer winding diameter
$F_{k-\epsilon}$	–	turbulent modelling function
$F_{k-\omega}$	–	turbulent modelling function
$\vec{f}$	N/m^3	source term inside Navier-Stokes equations
Gr	–	Grashof number
g_r	K	average winding to oil temperature gradient
h	J/kg	enthalpy
h_{init}	μm	initial cell height in boundary layer
h_{model}	m	model height

H	–	hot spot factor
I	A	electrical current
$\vec{I}$	–	identity tensor
K_{geom}	–	characteristic geometrical dimensionless property
k	J/kg	mass specific turbulent kinetic energy
L_{age}	h	product life time
L_0	m	characteristic length
L	m	length
M	–	discretization scheme
M	kg	mass
$\dot{m}$	kg/s	mass flow
N	–	number of winding turns
$\vec{n}$	–	normal vector
Nu	–	Nusselt number
P	W	power
P_{age}	1/°C	product lifetime parameter
$P_{loss,turn}$	W	power dissipation per winding turn
Pr	–	Prandtl number
p	Pa	pressure
$\Delta p_{hydraulic}$	Pa	hydraulic pressure loss
Δp_{stat}	Pa	static pressure difference
P_{total}	W	total losses in heated model
$\dot{Q}$	W	heat flow rate
$\dot{q}$	W/m^2	heat flux

Re	–	Reynolds number
r_{rad}	–	reflectivity
R_{th}	K/W	thermal resistance
T	K	thermodynamic temperature
t	s	time
ΔT_c	K	conductor temperature difference
U	J	inner energy
u	m/s	x-component of velocity in Cartesian coordinate system
$U_{channel,v}$	m	circumference of vertical cooling channel at inner winding diameter
v	m/s	velocity in Cartesian coordinate system
$\dot{W}$	W/m^3	volumetric power density
w_0	m/s	characteristic flow velocity
y^+	–	y-plus
x	m	x-component in the Cartesian coordinate system
y	m	y-component in the Cartesian coordinate system
z	m	z-component in the Cartesian coordinate system

Greek Symboles

Symbol	Dimension	Description
α	$W/(m^2.K)$	heat transfer coefficient
β	–	thermal expansion factor
$\Delta rel,i$	%	relative differences
μ	$Pa.s$	dynamic viscosity
μ_t	$Pa.s$	turbulent dynamic viscosity
ε_{rad}	–	emissivity
ε	$J/(kg.s)$	turbulent dissipation
γ	–	dimensionless factor for analytical assessments
λ	$W/(m.K)$	thermal conductivity
ν	m^2/s	kinematic viscosity
ρ	kg/m^3	density
σ_ϕ	–	turbulent Prandtl number
θ	°C	temperature in Celsius scale
$\vec{\tau}$	N/m^2	stress tensor
θ_0	°C	characteristic temperature
$\Delta\theta_0$	K	characteristic temperature difference
θ_{HST}	°C	hot spot temperature
$\Delta\theta_{oil}$	K	temperature difference between model inlet and outlet
τ_{ij}	s	Reynolds Stress Tensor
φ	rad	rotational dimension in circumferential winding direction
ω	$1/s$	specific dissipation rate

Constants

Symbol	Description
g	gravitational constant equal to 9.81 m/s^2
σ	Stefan-Boltzmann constant equal to $5.67 \times 10^{-8} \frac{\text{W}}{\text{m}^2.\text{K}^4}$

Abbreviations

AC	Alternating Current
CAD	Computer-Aided Design
CFD	Computational Fluid Dynamics
CNC	Computerized Numerical Control
CTC	Continuously Transposed Conductors
DC	Direct Current
GUI	Graphical User Interface
HTC	Heat Transfer Coefficient
HST	Hot Spot Temperature
HV	High Voltage
HWA	Hot Wire Anemometry
IEC	International Electrotechnical Commission
IEEE	Institute of Electrical and Electronics Engineers
LDV	Laser Doppler Velocimetry
OD	Oil Directed
OF	Oil Forced
ON	Oil Natural

P&I	Piping and Instrumentation (diagram)
PIV	Particle Image Velocimetry
PTFE	Polytetrafluoroethylene
PVDF	Polyvinylidene Fluoride
RANS	Reynolds-Averaged Navier-Stokes
RMS	Root Mean Square
SST	Shear-Stress-Transport
THNM	Thermal Hydraulic Network Model

Indices

2D	reference to results from 2D models
3D	reference to results from 3D models
Amb.	reference to ambient conditions
Cond.	reference to thermal conduction
Conv.	reference to thermal convection
f	reference to bulk flow
In	reference to inlet conditions
Measure	reference to measured values
Max	reference to measured values
Min	reference to minimum values
Rad	reference to thermal radiation
Rated	reference to rated specifications

1. Introduction

1.1 Importance of Thermal Modelling in Power Transformers

Since power transformers are the most valuable components in energy transmission, it is necessary to consider the significant parameters affecting their ageing. In this regard, the rapid increase of renewable energy resources in modern intelligent grid systems has increased the demand for power distribution systems.

The power rating and the service life of power transformers directly depend on the thermal aspects. Power systems connected to renewable sources typically experience various load ratings compared to conventional energy resources, leading to transformer overloading. Thus, it is always necessary for facilities to enhance the loading capacities of the transformers while maintaining their reliability and life expectation based on an appropriate criterion [1].

Unexpected thermal overloading or continuous thermal stresses can accelerate the degradation of insulation materials and reduce the breakdown voltage in transformers. The highest temperature within the insulation system is called the hot spot temperature, and winding hot spot temperature is a crucial factor in the ageing of the insulation domain. Two parameters affect the location of the hot spot and the highest temperature within the winding; first, the heat dissipated within the transformers due to electrical power losses. Second, the efficiency of the heat transfer mechanism and cooling characteristics resulting from the distribution of cooling liquids. Figure 1.1 show the overheating of the discs located at the top of the Low Voltage (LV) winding [1].

Figure 1.1: Overheating of the conductors at the top of the LV winding of the transformer [1].

Excessive heat generation in power transformers can cause overheated regions in the windings, and unexpected thermal stresses can cause early failures of the insulation materials and additional costs in the power network system.

Since the loading of transformers impacts the rate of heat losses, there is a need for profound knowledge about thermal stresses and temperature distribution. This necessitates a comprehensive study of heat sources within transformers and their cooling systems. Consequently, an optimized thermal design can achieve following international IEEE standards, mechanical structural techniques, and also environmental considerations. It also provides reliable components with the lowest costs. Various methods are employed to determine temperature distributions and oil flows within transformers.

Thermal sensors have been used in an experimental approach to directly monitor winding temperature with some limitations. The position of the hot spot temperature is mostly not predictable for the connection of thermal sensors. Therefore, measuring the hot spot temperature within in-service transformers is not possible. Moreover, the high costs of different experimental setups and monitoring systems motivate designers to use multiple modelling concepts, such as numerical approaches [2].

Numerical concepts are commonly used in transformer design due to their reasonable accuracy and ability to incorporate different boundaries and operational conditions for optimized designs. While IEC 60076-2 standards provide fundamental thermal considerations for power systems using empirical factors, accurate numerical calculations are necessary to achieve optimized models for various environmental conditions [2].

1.2 Objective and Motivation

The critical ageing process of power transformers in the electrical power network prompts the importance of conducting a comprehensive investigation of thermal performances. Due to the vital requirement of high-performance cooling systems within power transformers, significant numerical and experimental investigations have been developed. Presenting an optimized cooling model indicating lower thermal stresses with more effective thermal performances will be valuable. Thus, this study aims to perform thermal analyses of power transformers using CFD method with combination of experimental approaches to validate numerical investigations.

Furthermore, this work aims to optimize the cooling concepts of transformers by investigating temperature measurement and oil flow distribution. To achieve this, the temperature should be measured at two separate cooling modes, oil-directed (OD) and oil-natural cooling modes (ON). 3D CFD numerical models correspond to the investigated experimental designs should be validated by comparison with measured data.

The hot spot factor in the cooling of power transformers is a critical aspect of transformer design and operation, and its studying is essential for several reasons: Preventing Overheating: Transformers operate under various load conditions, and excessive heat generation can lead to overheating of critical components such as windings and insulations. The hot spot factor helps to predict the hot spot temperature within the transformer, which is often located in the winding, and ensures it remains within acceptable limits. Overheating can cause insulation degradation and reduced transformer life.

Maintaining Efficiency: The efficiency of a power transformer is directly linked to its operating temperature. Higher temperatures result in increased losses, reducing the efficiency of transformers. By studying the hot spot factor and hot spot temperature at different cooling modes within this study, engineers can consider only the most remarkable parameters to maintain the temperature of transformers at an efficient level. Determining Loading Capacity: Transformers have rated loading capacities, and the hot spot temperature is a critical factor in determining how close a transformer is operating to its capacity. By determining the hot spot factor and keeping hot spot temperature within defined limits, operators can make decisions about load adjustments and prevent overloading, which can lead to premature failure.

The standard hot spot factors, outlined in the IEC standards, may not always serve as a substitute for a comprehensive thermal analysis. Relying on a default hot spot factor can provide a rough estimation. It is crucial to calculate hot spot factor under various operational conditions. This approach directly addresses the critical question of how different parameters impact hot spot factor and hot spot temperature. Therefore, in this study, the hot spot factor for different cooling designs should be determined based on the measurements and CFD simulation results.

To pursuit of these objectives, the hot spot factor will be determined for a winding model under ON and OD cooling modes. Additionally, by utilizing a real distribution transformer, the hot spot factor during a heat run test will be determined. Two CFD models should be executed to determine the hot spot factor by using average temperature of conductors. Furthermore, geometrical winding designs will be undertaken to optimize the cooling concepts by evaluating the oil flow and temperature distribution. Various liquid materials should be used by illustrating oil flow distributions, and temperature distribution at different CFD designs. The impacts of power loss distributions should be identified by considering the position and temperature of the hot spot at different operational conditions. All these investigations aim to comprehensively address all the influential parameters. With this light, the outlines of the study are given in the next section.

1.3 Structure

This work is structured into six sections. Chapter 1 presents the introduction, the motivation behind the study, and outlines the main objectives. In chapter 2, the basic principles of power transformers are presented and fundamentals of both solid and fluid insulations are introduced. Additionally, basic transformer thermal modelling concepts, including no-load and load losses, are explained.

Chapter 2 also provides the reasons of heat losses in power transformers and offers a comprehensive understanding of fluid flow behaviour and heat transfer mechanisms. It also presents the background of this topic, specifically the state-of-the-art design and determines the most exhaustive studies. It includes experimental approaches and numerical methods, consists of thermal-hydraulic network model (THNM), equivalent circuit models, finite element, and finite differential models.

Next, chapter 3 details the analyses CFD numerical models, describing the discretization scheme linked to the solving procedures.

Chapter 4 discusses the parameters affecting the cooling behavior of the power transformers operating in the OD cooling mode. It shows that operational conditions indicating flow rate, liquid materials and oil inlet temperature significantly affect the cooling efficacy. Geometrical analysis is carried out to evaluate the thermal performance of different designs.

Furthermore, this chapter contributes valuable insights into the thermal behavior of biodegradable oils within the context of OD cooling mode. It concludes by highlighting the profound impacts of employing different oil liquids on both average winding temperature and the critical hot spot temperature.

Chapter 5 focuses on the transient thermal behavior of power transformers. Two models are collected to investigate the different operational scenarios under ON cooling mode. The transient temperature distribution is determined using a heat run test within a distribution transformer, providing essential insights into the dynamic performance of hot spot factor and hot spot temperature.

Finally, chapter 6 presents the conclusions and recommendations for future research directions. For those seeking a deeper understanding of the material properties within the solid and fluid domains, the material properties of solid and fluid domains are addressed in Appendix A.1 while more details about modelling the turbulent flow are presented in Appendix A.2.

2. Theoretical Principles

2.1 Power Transformer Technology

This chapter describes the technical details of power transformers and cooling conditions related to the scopes of investigation. The discussion in this chapter provides basic knowledge about the structure of power transformers to give a comprehensive view of thermal performance. Moreover, before analysing the thermal performance of the power transformers in power systems, it is essential to figure out the primary source of heat losses in the power transformers. Firstly, the structures of different transformers, particularly disc type power transformers are expressed. Following the winding design, the insulation and cooling liquids are explained.

Afterwards, the classification of the heat losses in power transformers is illustrated. After discussing the cooling concepts of the power transformers, the reference properties in thermal design with the calculation method of the hot spot factor are considered. The fundamentals of the momentum and energy equations are introduced, and state-of-the-art works in the thermal modelling of power transformers are reviewed. This chapter highlights the attempts to study the thermal characteristics of transformers.

2.1.1 Basic Principles of Power Transformers

2.1.1.1 Transformer Constructions

Before introducing power transformers, the basic design of transformers is presented. A simple two-winding transformer construction consists of winding being wound on a separate soft iron limb or core which provides the necessary magnetic circuit. A transformer construction provides a magnetic circuit, known more commonly as the transformer core, which is designed to provide a path for the magnetic field to flow around. Generally, the winding side which is connected to the voltage source and create magnetic flux called the primary winding. The other winding is called the secondary winding in which a voltage is induced as a result of induction. This magnetic path is necessary for the induction of voltage between windings. The two most common designs of the transformer constructions are classified into the core type and shell type transformers.

In the core type transformer, the primary and secondary windings are wound outside and surround the core ring. In the shell type transformer, the primary and secondary windings pass inside the steel magnetic core which forms a shell around the windings. In core type transformer construction, the windings are usually arranged concentrically around the core limb with the higher voltage at primary winding being wound over the lower voltage at secondary winding.

Figure 2.1 depicts the arrangement of different windings and cores by designing transformers. Transformer windings form another important part of transformer construction, because they are the main current-carrying conductors wound around the laminated sections of the core. The type of conductors which are used to carry the current in a transformer winding is either copper or aluminum. While aluminum conductor is lighter and generally less expensive than copper, a larger cross-sectional area of conductor must be used to carry the same amount of current as with copper so it is used mainly in larger power transformer applications [3].

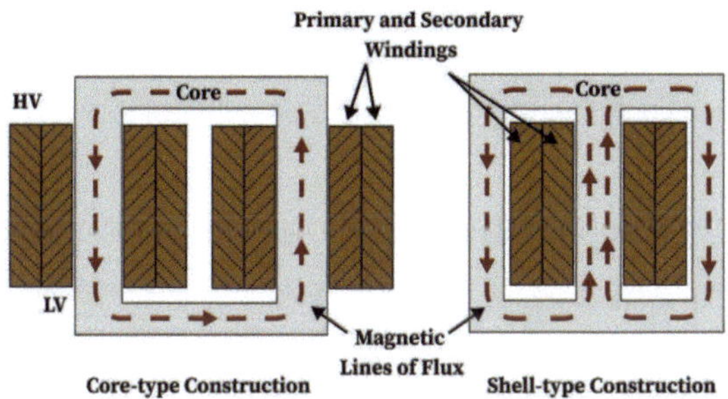

Figure 2.1: Cut-through of a limbs and cores in transformer designs [3].

2.1.1.2 Winding Type

At the core of the designs, each power transformer has at least two windings, specifically low voltage (LV) winding and high voltage (HV) winding. The tapping is usually connected to the HV winding in the case of a variable transformation ratio, and the corresponding turns are separated accordingly. Some high-rating power transformers are equipped with a third tertiary winding.

Apart from the LV and HV windings, the tertiary winding is connected in a delta formation. The tertiary winding assists in limitation of fault current in the event of a short-circuit from line to neutral.

Moreover, with an unbalanced load during the operation, the unbalancing in LV windings can be reduced by using tertiary winding. The tertiary winding applies auxiliary loading from HV loading at different voltage levels.

A variety of different types of windings have been used in power transformers through the years. Conductors can be wound in an upright, vertical orientation, as is necessary with larger, heavier conductors; or they can be wound horizontally and placed upright upon completion [3]. The type of winding depends on the transformer rating as well as the core construction. Some types of windings are commonly referred to "pancake" windings due to the arrangement of conductors into discs. However, the term most often refers to a coil type that is used almost exclusively in shell form transformers. The conductors are wound around a rectangular form, with the widest face of the conductor oriented either horizontally or vertically [3].

Helical windings are referred to as screw or spiral windings, with each term accurately characterizing the coil's construction. A helical winding consists of a few to more than 100 insulated strands wound in parallel continuously along the length of the cylinder, with spacers inserted between adjacent turns or discs and suitable transpositions included to minimize circulating currents between parallel strands [3].

Layer windings are among the simplest winding configurations, as they involve the insulated coated conductors being wound closely around the cylinder core with using spacers. Several layers can be wound on top of one another, with the layers separated by solid insulation, ducts, or a combination. Variations of this winding are often used for applications such as tap windings used in load-tap-changing transformers and for tertiary windings [3].

Disc-type windings are one of the most important winding types; a detailed top view of one phase, including the core and shields, is depicted in Figure 2.2. The layered type is usually used for the LV winding, and due to the low voltage rate of the low voltage winding, it is located close to the grounded core. It is wound in vertical layers (top-bottom, bottom-top) on a robust tube. High-strength insulating material such as synthetic resin-bonded paper is employed in the construction procedure [3].

A disc winding can involve a single strand or several strands of insulated conductors wound in a series of parallel discs of horizontal orientation, with the discs connected at either the inside or outside as a crossover point.

Each disc comprises multiple turns wound over other turns, with the crossovers alternating between inside and outside. Most windings of 25-kV class and above use in core form transformers are disc type [3]. Power transformer windings are wounded as concentric cylinders, enclosing the winding when the core provides the flux return path. Moreover, it is preferable to use disc type winding in HV winding [4]. It is possible to interleave discs near the HV lead exit for high voltage surges, and this method influences stray capacitances between discs. In this light, it is possible to shape the distribution of surge voltage along the winding height with this method. Therefore, to reduce eddy current losses, dividing the conductors in the direction perpendicular to the leakage direction (the radial direction) is necessary.

The axial sticks made from the pressboard will be located to provide axial cooling channels for the transformer oil. The radial cooling channels are designed by using a spacer between the discs. Pressboard cylinders, separated by oil gaps, are used to insulate LV and HV windings from each other. Using pressboards also brings a high dielectric strength barrier. Moreover, axial sticks provide sufficient firm supporters for HV windings during the manufacturing process of disc type winding. The horizontal layers in the disc type winding are discs, and the winding sequence is alternately in-out, out-in, and so on. A disc includes the turns that touch each other or are separated by pressboard spacers. Figure 2.3 shows a disc type winding design during the manufacturing process. Moreover, there is a gap between each disc for the radial oil channel, and the conductors are wrapped using insulation.

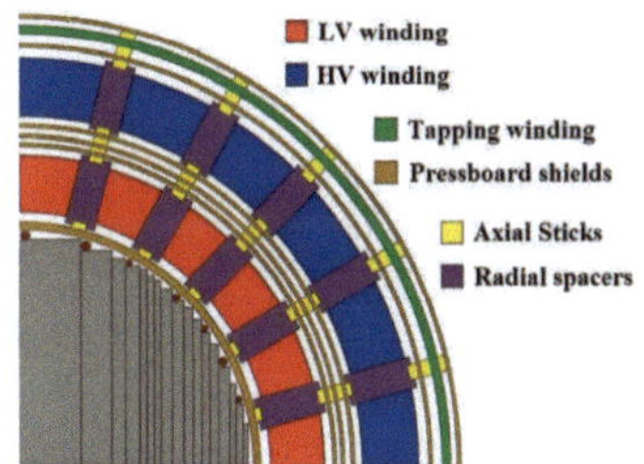

Figure 2.2: Cut-through a 5 limb 3 phase core, top view [3]

Figure 2.3: Construction process of disc-type winding in power transformer, including radial channels [4].

The gap between turns provides axial cooling channels to enhance cooling conditions. In this regard, the radial spacers offer not only radial cooling channels but can keep the winding structure firmly in the axial direction.

2.1.1.3 Continuously Transposed Conductors (CTC) Winding Type

High-conductivity copper is usually used for large and medium power transformers during manufacturing. The rectangular-shaped conductor is the most popular used in windings, offering the highest space in the windings. The latter defines the ratio between copper cross-section and total winding cross-section. The space factor decreases using circular cross-section wires, as many gaps exist between the turns [4-5]. Concerning the mechanical design, rectangular wires offer more stability than round wires, so circular cross-section wires are limited to the designs for distribution transformer sizes.

Rectangular cross-section wires can be classified into single-stranded flat conductors, continuously transposed conductors (CTC), and bunched conductors. Figure 2.4 presents schematic examples of different winding types. Conductors are usually coated with polyvinyl acetate material with a 50-70 micrometers thickness range to insulate them from each other. The insulated conductors are wrapped with more insulation materials between the winding turns. There are different types of conductors, and their applications are shown in Table 2.1. Single-stranded conductors and bunched conductors can be insulated by enamel or paper. Sometimes, the strands are insulated by enamel, and the whole bundle is wrapped in paper.

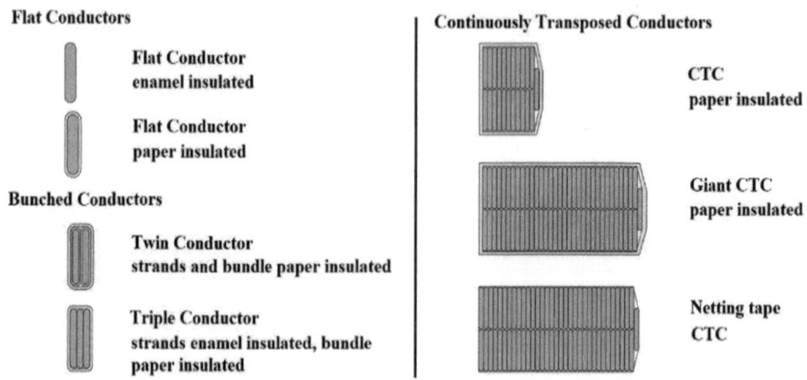

Figure 2.4: Conductors for transformer windings [4].

Some conductors have a paper cover around the entire bundle, while others have individual paper covers around each strand. CTCs consist of several strands that are insulated by enamel and stacked together. They usually have a paper cover around the whole bundle. CTCs are becoming popular because they reduce eddy current losses by using small strands and transposing them. This way, the loop voltages are balanced by making each strand go through the same average stray magnetic field in a winding turn [6]. CTCs also allow for a smaller core and transformer size and lower construction costs because they use more copper and less insulating materials in the core window.

Table 2.1: Different types of conductors and their applications [4].

Arrangement	Application	Picture
CORDEX only with protection paper	LV or HV winding	
CORDEX 1 with 2 layers of paper at the bottom in radial dimension	LV winding split CTC in two due to the big radial dimension	
CORDEX 2 with 2 layers of paper in the axial dimension	Layer winding	
CORDEX 3 with aramid paper in radial dimension	HV winding	

2.1.1.4 Solid Insulation

According to Kulkarni and Khaparde [7], electrical insulations in power transformers can be classified into minor and major insulation structures. The classification of major and minor insulation refers to the level of dielectric strength.

The dielectric strength reflects the ability of the material to withstand an electrical field without losing the electrically insulating qualities and is defined in units of kV per mm. The primary insulation guarantees electrical isolation between the windings, the windings and all grounded components (i.e., the core) and between all ungrounded leads and grounded components. The major insulation system includes barriers, spacers, and clamps, while the minor insulation consists of winding insulation. Generally, the combination of insulation oil and oil-impregnated cellulose-based materials is utilized when there are no delicate environmental or high safety standards due to its cost-effective electrical insulation capabilities.

Cellulosic transformer insulation has three functions. First, it acts as a dielectric by storing electrical charge when the transformer is energized. Second, it fulfils a mechanical function by supporting the windings. Third, it contributes to a better thermal performance of the transformers by creating cooling channels for the oil [8]. Figures 2.5 and 2.6 show different types of insulations around the conductors. The cellulose fibers are oriented to obtain higher tensile strength and stiffness by production. The number of layers around the conductors relates to the desired dielectric strengths.

The increased thickness of the dielectric material provides higher withstand in an electrical field. The thermal conductivity of the insulation materials in the oil-impregnated condition is assumed 0.1- 0.2 $W/(m.k)$ [9-10]. Recently, paperless CTCs are commonly used for the conductors in LV windings of large power transformers. The thermal performance of the conductors without paper is higher than that of the covered conductors due to the lower thermal resistance between oil and conductors.

Figure 2.5: CTCs covered by insulation papers [8].   Figure 2.6: Layers of different insulations [8].

Standards IEC 60554 [11], IEC 60641 [12], and IEC 60763 [13] present comprehensive studies on the scope of insulation papers and pressboard materials, including adhesives in washers, commonly made of cardboard material, can be installed regularly to ensure a sufficient oil flow distribution within the radial cooling channels. They force the oil flow through the radial channels inside the disc type winding. The washers divide the winding into passes, which provide better cooling conditions. This method is not limited to the disc type of winding.

It is also used in helical winding designs for similar purposes. The number of washers and discs per pass depends on the design structure and can be changed for optimization. Washers are employed in the disc type winding models to create zig-zag cooling pattern.

This study investigates the zig-zag arrangement of the disc type windings to analyse the thermal performances. Another popular design for the winding is to use additional vertical cooling channel. Instead of using thick solid insulations between conductors in the disc type design, the insulation is replaced with a narrow vertical cooling channel. The new vertical oil channel will distribute the oil more effectively along the winding without changing the dimensions of the active parts.

2.1.1.5 Liquid Insulation

The first generations of power transformers only used mineral oil for cooling purposes due to the high dielectric strength and the high capability to transfer generated heat losses.

Mineral oils also perform well in applications such as load tap changers where arcing occurs in the insulating liquid. Generally, mineral-insulating liquids have lower costs compared to other liquids. Natural esters are environmentally friendly dielectric liquids that are biodegradable. Natural ester oils are derived from corn, rapeseed, soybean, and sunflower crops. Besides being an advantage for spill containment and material handling, these liquids have some performance advantages over mineral oil.

For indoor applications where flammability is a critical concern, ester oils are usually employed due to higher flash and fire points with a high capacity for moisture saturation. Synthetic ester liquids are available with similar characteristics and can increase low-temperature performance. The types of ester oils are formulated to prevent oxidation when exposed to air, as in free-breathing applications [14].

Appendix A.1 provides the material properties of the mineral oil and natural ester oil which are considered in this work.

2.1.2 Transformer Losses

An essential step before evaluating the thermal performance within the winding is determining the amplitude and the distribution of the electromagnetic losses [7]. Each material shows resistance when an electrical current flow through it. Different rates of heat energy can be generated depending on the electrical resistance and current density through the material. Meanwhile, despite the generated heat losses in the conductors due to the current density, an in-service power transformer will suffer other losses, including heat losses in the core, tank, and other iron parts. These losses are generated due to changes in the magnetic field within the power transformer [7]. Thus, the source of losses should also be known and explained to analyse the impact of the heat losses for the cooling consideration. The classification of total losses within the power transformers is illustrated in Figure 2.7.

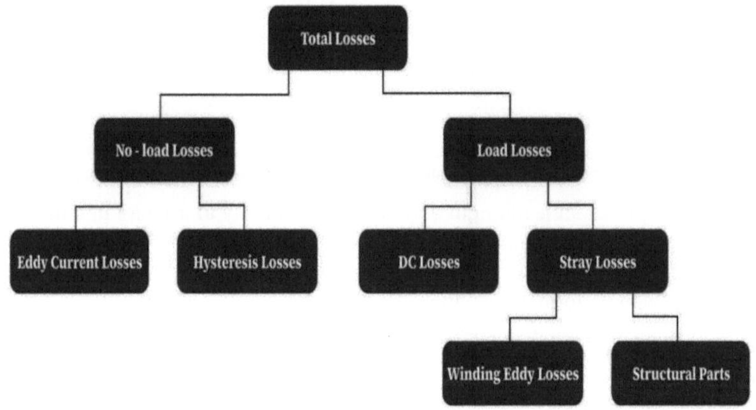

Figure 2.7: Transformer losses classification [15].

Transformer losses are broadly classified as no-load and load losses [7]. No-load losses depend primarily on the voltage and frequency, so under operational conditions, they vary only slightly with system variations. Load losses are, in turn, broadly classified as Ohmic losses due to Joule heating produced by the current flow through the conductors and as stray losses. Both losses depend on the applied loading rate of the transformers.

Stray losses are produced due to the accumulation of additional losses, such as tank walls, clamps, or bracing structures usually experienced by the transformers [16]. Stray losses indicate the winding eddy losses and losses due to leakage flux entering the internal metallic structures. The significant increase in the eddy losses for the top discs is attributed to the considerable contribution of the radial leakage flux. Figure 2.8 shows a schematic example of magnetic leakage flux with the magnified radial leakage flux. In contrast to the ohmic losses, which are almost constant within the conductors, the stray losses are distributed non-uniformly along the winding.

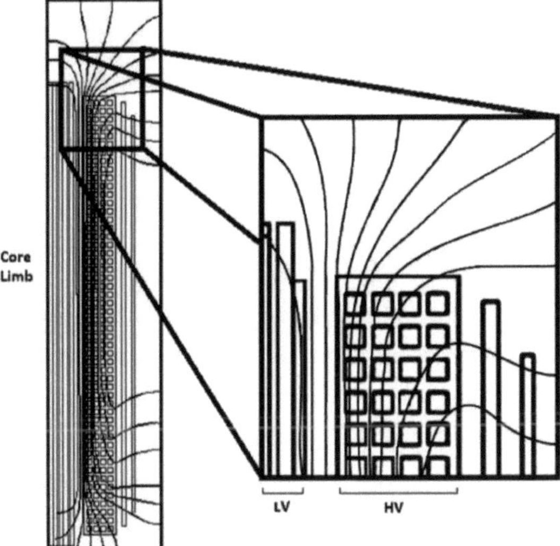

Figure 2.8: Schematic view of leakage flux of the windings and the magnified view of the large radial leakage flux components over the top discs [1].

2.1.2.1 No-Load Losses

No-load losses occur when a transformer is energized with its rated voltage at one set of terminals, and the other sets are left open to ensure no current flows. In this scenario, total flux arises in the core, and the windings have only the necessary exiting current flows. These losses are predominately eddy currents produced by the time-varying flux in the core and the hysteresis losses [17].

The magnitude of the no-load losses depends on various factors such as core design, quality of the core plate materials, induction, and core mass [18].

Keeping the hysteresis and eddy current losses under a desired level is essential for efficient operations.

Hysteresis losses are produced due to the elementary magnets in the material aligning with the alternating magnetic field. The core material becomes magnetized by applying a magnetic field through a core. Consequently, the layers within the core must align themselves with the external magnetic field. The molecular friction in the core material produces hysteresis losses by redirecting the magnetic field. A fluctuating magnetic field emerges when an alternating-current source energizes the transformer. The induced voltage leads to current flows through the core, which generates heat energy. The undesired currents are called eddy currents. The core of the transformers is laminated during the construction to reduce the losses created by eddy currents. In this case, the eddy current losses can be minimized as the insulated laminations have no path for currents.

2.1.2.2 Load Losses

Load or short-circuit losses occur when the output is connected to a load to allow the current to flow through the transformer from input to output terminals. Although core losses also occur in this scenario, they are not considered part of the load losses in the definition [17]. The output terminals are grounded when measuring the load losses, and only a small impedance-related voltage is necessary to produce the desired full load current [17]. In this case, the core losses are small and can be neglected during the measurements of the losses due to the small core flux.

2.1.3 Transformer Cooling Systems

The previous section considers the primary source of losses in power transformers. Generally, heat energy is generated due to the electrical losses inside the system. The main concept of cooling in the power transformers is focused on dissipating heat losses. The active part of transformers is made of the elements that are in contact with the voltage and the current and are mainly composed of windings, core, and tap changer bushings.

Hence, in the oil-immersed transformers, the cooling oil will be used to transport the heat via convection method. Meanwhile, as defined in standard IEC 60076-2 [19], the first letter is utilized to identify the cooling medium. For example, the first letter O signifies the cooling oil, whereas the second letter (F, N, D) shows the cooling modes.

Radiators are used in the cooling mode ONAN (Oil Natural Air Natural) and ONAF (Oil Natural Air Force) for heat exchange. The use of pumps in cooling systems allows for the achievement of forced convection.

Oil flow may be increased in the oil force (OF) mode due to the external force of the pump within the winding channels and the transport of heat energy to the radiators. Moreover, oil-directed (OD) cooling mode refers to the forced oil convection inside the cooling accessories and windings, directing the oil through the channels via washers.

In the natural cooling mode (ON), the oil flows upward due to the thermal driving force, called buoyancy, in the fluid domain. The temperature of the oil will be lowered in the radiators and moves downwards to the inlet of the active part. It moves freely through the pipelines due to pressure differences, allowing the oil to circulate even without any external force on the fluid.

In medium-size or distribution transformers, the outer fins dissipate heat energy to the environment. Furthermore, additional cooling fans are provided for external radiators. The radiators can improve the cooling efficiency to enhance the thermal driving force. Principle illustrations for the internal cooling concepts are given in Figure 2.9.

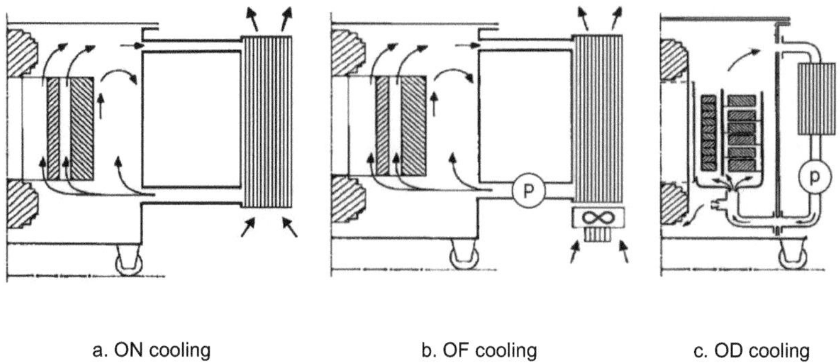

Figure 2.9: Illustration of principle concepts for internal cooling.

Since this work focuses on oil-directed and natural cooling modes, the investigated concepts can be identified with OD and ON cooling modes [19].

2.1.4 Hot Spot Determination

The loading conditions primarily determine the thermal behavior of windings, which directly depends on the solid insulations. Solid insulations refer to pressboard and beech wood with materials such as Kraft, Crepe, and thermally upgraded paper. The cellulose-based insulation papers are high-polymer carbohydrate chains usually made from glucose-related molecules. These materials experience depolymerization and ageing over time due to oxidation and acidosis. Decreasing papers over time is directly connected to the temperature range. Higher temperatures accelerate the ageing rate of insulation papers, leading to a shorter solid insulation lifetime.

Since the temperature distribution is not uniform in transformers, the component which is experience the highest thermal stress will be exposed to the greatest deterioration. Meanwhile, in ageing studies, it is usual to consider the significant effects of hot spot temperature. Aging of power transformers, affects the reliability of the system and is an important criterion in transformer's asset management. As the age of the transformers increases, the capabilities of transformers to withstand severe events such as short circuit faults decrease. It results in increase in probability of failure during operation.

By online monitoring the thermal behaviour of power transformers and controlling the hot spot temperature, utility owners can have an estimation on transformers ageing rate. It leads to have an appropriate time for maintenance process.

The ageing rate can be expressed as a reaction rate constant L_{age}.

According to Montsinger [20], Arrhenius reaction rate equation or Dakin relationship can be applied to the approximation within limited temperature ranges by

$$L_{age} = D_{age} \cdot e^{\frac{B}{\theta_{HST}+273}} \qquad \qquad \text{Equation 2.1}$$

where D_{age} refers to the lifetime,

and B is the specific constant with the unit (1 / °C).

θ_{HST} is the hot spot temperature (°C).

According to IEC 60085, an 8 K increase in the insulation material in the thermal class A insulation material leads to a doubling of the ageing rate in the temperature range of 80 °C to 140 °C [21]. This statement is accepted by transformer manufacturers around the world [7].

The lifetime of the thermally upgraded insulation papers is also referenced in the contributions [22]. It is evident that the lifetime of insulations depends on the highest temperature of the insulation material, even if the average winding temperature is moderate. Therefore, overheated locations can cause serious failures during operations.

Prior to the design phase, it is imperative to establish the operating conditions in order to compute the hot spot temperature. By applying the value of hot spot temperature in the estimations, the lifetime of the power transformers under a specified load is predictable. In this contribution, defining the reference properties for comparing the thermal performances at different boundary settings is essential.

Therefore, Figure 2.10 depicts the diagram of the idealized temperature distribution along the transformer height according to standard IEC 60076-2: 2011 [23].

This diagram includes some specific simplifications, which are contributed below, such as:

- It is noticeable to assume a linear oil temperature rise from the bottom to the top inside the winding. Furthermore, at any position along the winding height, the temperature increase within the conductors is parallel to the rise in the oil temperature. Therefore, the difference between the average winding temperature (Point F) and the average oil temperature (Point C) can be expressed by the term g_r.

- The hot spot temperature (Point P) is higher than the temperature of the conductors at the top of the winding because of local eddy currents.

    The difference between the hot spot temperature and top oil temperature (point A) of the winding is equal to $H \times g_r$.

- Measured values are identified by a solid triangle (▲), whereas the calculated values are identified by a solid circle (●).

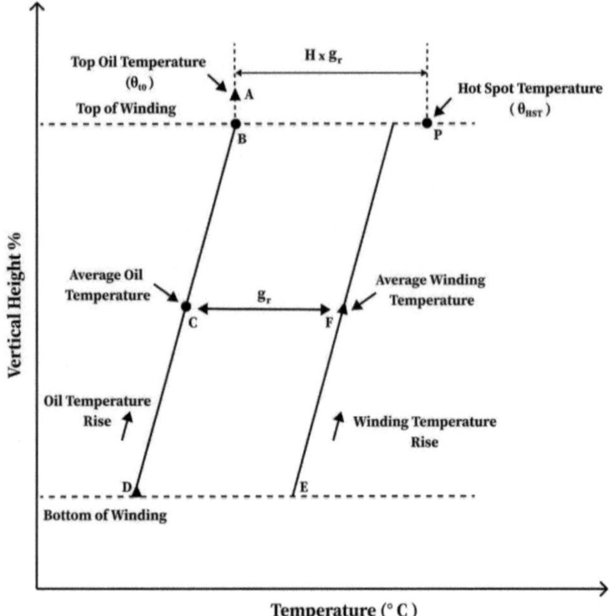

Figure 2.10: Idealized temperature rise within the oil and the winding from bottom to top along the winding height, according to IEC 60076-2:2011 [19].

According to the thermal diagram in Figure 2.10, the location of the hot spot temperature is assumed at the top height of the winding. However, in most power transformers, the hot spot temperature is located elsewhere; the hot spot can be located at the lower turns of the winding because of the magnetic flux leakage and the different oil distribution in the cooling channels.

The dimensionless constant factor H is defined in Figure 2.10, and this factor represents the increase of the average winding gradient due to the local increase of additional loss and the variations in the liquid flow. Based on the thermal diagram in Figure 2.10, the hot spot factor is assumed to have a constant value, although it depends on different loading conditions. The loading condition determines the type of cooling mode, the design of the winding cooling channels, the proportion of the oil in the cooling channels, and the ratio of the additional losses.

Therefore, the hot spot factor may vary in different power transformers. By using the hot spot factor, the hot spot temperature can be calculated [24-25].

The hot spot temperature rise can be estimated by measured temperature rise from the temperature rise test.

$\Delta\theta_{HST}$ is given by equation 2.2.

$$\Delta\theta_{HST} = \Delta\theta_{to} + H \times g_r \qquad\qquad \text{Equation 2.2}$$

The average thermal gradient between each winding and liquid along the limb is taken as the difference between the average winding temperature rise and average liquid temperature rise. Consequently, the dimensionless hot spot factor H is applied below.

$$H = \frac{\text{Hot spot temperature} \; - \; \text{Temperature of top oil}}{\text{Average winding temperature rise} \; - \; \text{Average oil temperature rise}}$$

In IEC 60076-2:2011 [19], the H factor is defined through the Q factor and the S factor as $H = Q \times S$. The Q factor is carried out from the loss distribution within the transformer. It is a dimensionless factor as a ratio of two losses, and in cylindrical coordinates is defined according to

$$Q = \frac{q(r, z, \phi, T)}{q_{ave}} \qquad\qquad \text{Equation 2.3}$$

where $q\,(r, z, \varphi, T)$ is the local loss density at a location (W/m^3), r is the radial position, φ is the angle in circumferential position, z is the axial position, T is the local temperature at (r, z, φ) (K) and q_{ave} is the average loss of the winding at an average temperature (W/m^3).

Figure 2.11 shows the range of the Q factor which is directly depends on the transformer rating and on the copper winding strand width [19].

The S factor depends on the cooling efficiency of the winding. S factor is considered 1.0 for disc type windings without diverting washers, however the S factor can be changed to 1.2 in different systems, therefore the rate of the hot spot factor is changed from 1 to 1.8 by different ratings [19].

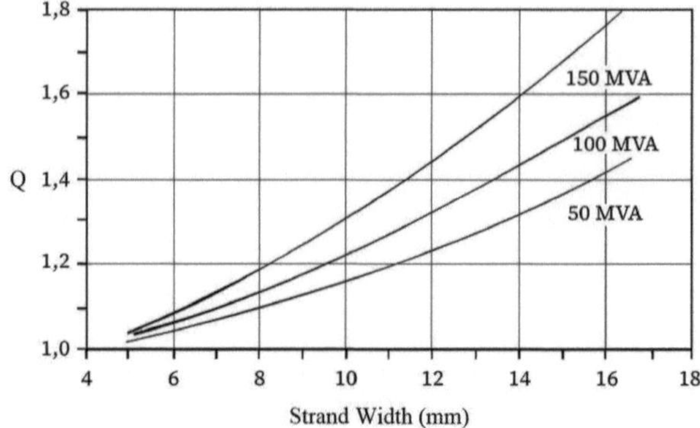

Figure 2.11: Range of the Q factor for different conductor strand width [19].

During thermal monitoring investigations, the hot spot temperature can be calculated using top oil temperature and load current or detected using different sensors mounted in power transformers [17-18]. However, it is usually not possible to measure the exact value of hot spot temperature during monitoring and measurement experiences because of the following reasons:

- It is impossible to evaluate the dependency of the factors affecting on hot spot temperature of a transformer only by measurements.

- Due to the various operating conditions of transformers, including the cooling condition and the loading, the position of the hot spot temperature cannot be determined through measurement.

- Installation of temperature sensors can only be performed at the production stage. Retrofitting the in-service power transformers is not possible.

The hot spot is generally expected to be located at the upper region of the winding due to the additional local eddy current losses. Furthermore, the variation of the liquid ducts at the top of the winding has an effect on the hot spot temperature. Therefore, according to IEC 60076-7 [23], hot spot factor is used to study the quality of cooling within windings. Non-uniform heat loss distribution and various cooling modes are the main parameters for different hot spot factors. Meanwhile, in this study, hot spot factors are determined at different operational conditions during transient ON cooling mode. This is a remarkable parameter to evaluate the thermal behaviour of power transformers.

2.2 Fundamentals of Heat Transfer and Fluid Flow

This section describes the fundamental principles most pertinent to the topic studied. First, it provides an overview of the theory underlying heat transfer before discussing the fundamentals of the governing equations in the fluid flow regime. The fundamentals of the momentum equation and energy equation are introduced. The modelling methods of turbulence fluid flow are explained, and the equations are explicitly considered in Appendix A2.

Regarding numerical investigations, the boundary settings are referenced in the following chapters to obtain primary conditions for the numerical approaches. The temperature field is determined by the solution of the energy equation, whereas the velocity distribution is resulted by solving the momentum equations. The mass conservation equation ensures that the fluid entering a control volume also leaves the control volume. The equations are used in 3D by Newtonian fluid and the liquid properties are prescribed as a function of temperature.

2.2.1 Heat Transfer

According to Incropera [26], the heat transfer mechanism is due to the temperature difference between the two domains. Higher temperatures, according to the second law of thermodynamics, determine the direction of the heat transfer. For energy transmission, the heat transfer theory stipulates three different physical principles. Generally, heat radiation, heat convection, and heat conduction are the main concepts of heat transfer. This section provides a brief overview of these fundamental concepts in heat transfer and the theory of each calculation method.

Baehr and Kabelac [27] and Boles [28] should be consulted explicitly regarding heat transmission from a thermodynamic perspective.

2.2.1.1 Heat Conduction

Heat conduction is a typical physical process for transferring heat that exits in all heat transfer processes. It is based on molecular, atomic, or subatomic particle interactions such as collisions, diffusions, or fusion. The heat flux density $\dot{q}$ (heat flux per unit area) can be estimated by considering the thermal conductivity k of the material under consideration and using the temperature gradient ∂T between two materials.

Considering Cartesian coordinates, according to Fourier's law of heat conduction, the relation between the heat flux and the local temperature is described in equation 2.4.

$$\dot{q} = -k[(\tfrac{\partial T}{\partial x})^2 + (\tfrac{\partial T}{\partial y})^2 + (\tfrac{\partial T}{\partial z})^2] \text{ with } k = k[T, P]$$

Equation 2.4

The second law of thermodynamics is reflected in the minus sign of the equation, which provides a heat flow in the direction of falling temperature levels. According to Baehr and Stephan [29], the pressure sensitivity of the thermal conductivity for solids and liquids may be neglected.

Conduction is the predominant heat dissipation method in solid domains, which becomes less prominent in the oil medium. As this work uses the transient heat conduction process, the first law of thermodynamics will be used to compute the transient temperature field inside incompressible solids while neglecting kinetic and gravitational potential energy changes. This accords to

$$\frac{\partial U}{\partial t} = P + \dot{Q} \quad \text{with}$$

$$\dot{Q} = -\int_{(A)} \dot{q}\vec{n}\, dA = -\int_{(V)} \mathrm{div}\dot{q}\, dV,$$

Equation 2.5

$$\frac{\partial U}{\partial t} = \int_{(V)} \rho\, c \frac{\partial T}{\partial t}\, dV \,,\, P = \int_{(V)} \dot{W}\, dV$$

The upper right-side section uses the divergence theorem to convert the surface integral with the normal vector, estimated based on the heat capacity of the material under consideration. The distribution of volumetric power density distribution $\dot{W}$ is used to determine the power source P.

The temperature independence of material properties $\vec{n}$ can be carried out accordingly by combining equation 2.4 and the specified components of equation 2.5.

$$\frac{\partial T}{\partial t} = \frac{k}{\rho c}\, \mathrm{div}\vec{\nabla} T + \frac{\dot{W}}{\rho c} = \frac{k}{\rho c} \cdot \left(\frac{\partial^2 T}{\partial x^2} + \frac{\partial^2 T}{\partial y^2} + \frac{\partial^2 T}{\partial z^2}\right) + \frac{\dot{W}}{\rho c}$$

Equation 2.6

In a nutshell, equation 2.6 provides the governing equation for determining the temperature distribution and considers the mechanism of conducting heat in the solid domains.

2.2.1.2 Heat Convection

Heat convection is the most important mechanism of heat transfer in fluids. This process is based on the fluid flow movement, which transfers the generated heat within the scope of cooling or heating. Based on the rate of the driving force, heat convection can be classified as natural or forced convection. If the fluid flow movement emerges due to external forces such as pumps, fans, etc., the heat convection represents a forced heat convection.

On the other hand, the driving force is created naturally without any externally driven flow, like pumps or fans, due to the fluid density changes that depend on the temperature differences within the oil domain. Fluid elements with a lower density generate a force in the opposite direction of the gravitational field. This process is known as buoyancy, and the resulting flow produces the natural convection heat transfer.

The heat transmission from a solid surface to the surrounding fluid is of primary relevance in thermal investigations. Regarding convection, the adjacent fluid to walls, known as the boundary layer, mainly affects the amount of the transferred heat. Fluids do not slide on walls except for gases at very low pressures. Because of this, the fluid velocity is zero at walls, and its temperature is considered the same as the adjusted solid domain.

As a result, the local flow pattern significantly influences local heat transmission. In the complex field, convective heat transfer is coupled to a specific geometrical arrangement and flow characteristics, which can be estimated employing empirical correlations. Examples of these correlations can be found in VDI-Wärmeatlas [30], which provides the exact calculation of the heat transfer coefficient in the different thermal scenarios. The convective heat transfer coefficient results from equation 2.7.

$$h = \frac{\dot{q}}{(T_w - T_f)}$$
Equation 2.7

where $\dot{q}$ represents the heat flux density and

$(T_w - T_f)$ defines the temperature difference between the fluid temperature and boundary surface temperature. In complex mediums, heat transfer refers to dimensionless numbers, which allows for comparing boundary conditions with other geometrical designs. At the interface between solid and fluid, heat conduction occurs during heat transmission due to the non-slip conditions at the boundary surface.

The heat flux density at the wall can be calculated using equation 2.4 by applying y-direction, normal to the solid surface. Finally, applying to equation 2.7 results the following equation.

$$h = -k \frac{\left(\frac{\partial T}{\partial y}\right)_w}{(T_w - T_f)} \qquad\qquad \text{Equation 2.8}$$

Equation 2.9 is used to create dimensionless geometric term y^* which is defined as

$$y^* = \frac{y}{L_0} \qquad\qquad \text{Equation 2.9}$$

L_0 refers to the characteristic's length in the domain.

Applying the given definitions of equation 2.9 in equation 2.8 provides

$$h = -\frac{k_f}{L_0} \cdot \frac{\left(\frac{\partial T^*}{\partial y^*}\right)_w}{T_w^* - T_f^*} \qquad\qquad \text{Equation 2.10}$$

According to [28], the dimensionless Nusselt number is defined from equation 2.11 with

$$Nu = \frac{hL_0}{k_f} = -\frac{\left(\frac{\partial T^*}{\partial y^*}\right)_w}{T_w^* - T_f^*} \qquad\qquad \text{Equation 2.11}$$

The heat transfer process directly depends on the temperature difference between two mediums. The temperature distribution within the fluid depends on the fluid flow velocity. In oil directed cooling mode, the inlet velocity can be set even at boundary conditions or obtained with material properties. It can be impacted by thermal and hydraulic parameters, including the fluid's density ρ, viscosity μ, thermal conductivity k and specific heat capacity, c_p. During the numerical calculation, three other essential dimensionless numbers are associated with defining the heat transfer analyses. These numbers characterize the fluid flow regime from the thermal and hydraulic points of view.

These dimensionless numbers are defined with the Reynolds number Re, the Grashof number, Gr , and the Prandtl number, Pr in equation 2.12.

$$Re = \frac{\rho \cdot V \cdot L_0}{\mu}, \qquad Pr = \frac{C_p \cdot \mu}{k}, \qquad Gr = \frac{g \cdot \beta \cdot \nabla T \cdot L_0^3}{v_f^2} \qquad \text{Equation 2.12}$$

The Reynolds number can be used to identify the regime of the flow. It indicates the ratio of inertial forces to viscous forces. When Re is very high, the inertial force is much higher than the viscous force. Then, the inertial force becomes the dominant force, which usually leads to turbulent flow; when Re is low, the laminar flow emerges within the entire domain of the fluid flow.

The Grashof number mainly determines the fluid pattern under a specific temperature. It can be expressed as the ratio of buoyancy to the viscous force. Thus, it is applied to describe the natural convection of the fluid flow. Oil flow is derived from the buoyancy generated due to the density variation as a function of the temperature. Usually, a high Grashof number of fluids indicates an increased capability of natural convection. The term delta T is the temperature gradient between the surface temperature and the bulk temperature. v_f is the kinematic viscosity, and β is the volumetric thermal expansion coefficient at the bulk flow temperature. The average heat transfer coefficient on the solid surface h_m can be determined with the respective Nu_m. It can be calculated using empirical equations with using the terms in the equation below.

$$Nu_m = f\left(Re, Pr, K_{geom}\right), \qquad \text{for forced convection and}$$

Equation 2.13

$$Nu_m = f\left(Gr, Pr, K_{geom}\right), \qquad \text{for natural convection.}$$

Equation 2.13 includes a new term K_{geom}, which represents the geometrical properties of the designed model.

For example, the Nusselt number inside the pipeline has $K_{geom} = {}^{L}\!/_{d}$

with the geometrical design diameter d and length L.

VDI-Wärmeatlas [30] can be referred to in different designs and applications. The Prandtl Number is the ratio of fluid viscosity and thermal diffusivity, and it shows the relation between the hydraulic and thermal properties of the fluid. The high Prandtl number in the oil shows the significant effect of viscosity over thermal diffusivity.

A buoyancy force is the net effect of body forces acting on a fluid element, in which there are density gradients.

For the natural convection model, the density gradient is induced by a temperature gradient and the body force is applied due to the gravitational field, acting in the negative vertical direction.

2.2.1.3 Heat Radiation

During thermal radiation, the thermal energy of a solid or non-solid body is converted into electromagnetic radiation energy. Generally, every component emits electromagnetic radiation, depending on its temperature, which is passed on to its surroundings. If the radiation takes up on a surface, a part of it will be reflected or transmitted, and the rest will be absorbed on its surface.

The transmission of energy by electromagnetic radiation can travel through a vacuum domain. Thermal radiation on or from non-transparent bodies can be considered a two-dimensional heat source or sink. The radiant energy is converted into thermal energy within a few micrometers below the surface [31]. The heat flux density of actual surfaces can be calculated with the Stefan-Boltzmann constant σ, according to [31].

$$\dot{q}_{rad} = \sigma. T^4 \quad \text{with} \quad \sigma = 5.67 \times 10^{-8} \ \frac{W}{m^2 . K^4} \qquad \text{Equation 2.14}$$

The maximum radiation rate can be achieved when the idealized black body is applicable. Considering the emission of the different materials, electromagnetic radiation is determined by emissivity $\varepsilon_{rad}(T)$.

This coefficient depends on the surface temperature and the material property at the surface of the material. Depending on the thermodynamic temperature of the emitting body, the radiated energy is dispersed over a wide range of wavelengths. The wavelength spectrum of the incoming radiation also affects the ability to absorb, reflect, and transmit energy. Considering a black body within the ambient temperature of $T_{amb.}$ results, the heat flux radiation density of

$$\dot{q}_{rad} = \sigma(\varepsilon_{rad} . T^4 - \alpha_{rad} . T^4_{amb.}) \qquad \text{Equation 2.15}$$

The application of equation 2.15 can be challenging because it accounts for absorptivity, which is known to depend on the wavelength spectrum of incoming radiation. Moreover, the necessary measurements for material absorptivity α_{rad} is typically unknown and hard to undertake.

Thus, for a simpler estimation of the impact of heat radiation on the overall heat transfer of a given technical application, some simplifications are assumed to equation 2.16. Assuming a grey body permits to have ε_{rad} equal to α_{rad}.

Moreover, the assumption of neglecting the dependency on the wavelength spectrum provides a simple computation of the heat flux density $\dot{q}^*_{rad}$ according to

$$\dot{q}^*_{rad} = \sigma\varepsilon_{rad}(T^4 - T^4_{amb.}) \qquad \text{Equation 2.16}$$

For comparison, $\dot{q}^*_{rad}$ with the heat flux density connected to convection, a radiation heat transfer coefficient α_{rad} can be achieved from equation 2.17.

$$\alpha_{rad} = \sigma\varepsilon_{rad}\,\frac{T^4 - T^4_{amb.}}{T - T_{amb.}} \qquad\qquad \text{Equation 2.17}$$

Equation 2.17 can be used to compare the influence of thermal radiation and the heat transfer mechanism. In this study, the difference between the usual operating temperature of conductors and cooling oil is approximately 20 K, and conductor temperatures are mostly kept at less than 120 °C. In this light, heat radiation between the conductors and the cooling oil can be substantially simplified by assuming the conductors and cooling oil are black bodies at bulk flow temperature.

This implies ignoring the boundary layer between them and assuming the cooling oil entirely absorbs all maximum heat generated within the conductors. In this study, two different models of fluid flow regimes are considered. The current investigation can neglect the radiation rate compared to the convection heat transfer by both forced and natural fluid flow.

In this work, the comparison between the radiation heat transfer and the convection heat transfer shows the neglectable amount of heat radiation in force and natural cooling modes. Furthermore, as described in Chapter 5, the experimental setup is fully covered by a thick layer of glass wool to maintain the entire of the oil volume under neglectable rate of heat radiation. Therefore, heat convection and conduction are important mechanism of heat transfer and will be considered during CFD investigations. The role of heat radiation has also been thoroughly described by Incropera [26].

2.2.2 Fluid Flow

The previous section has described the significant link between the fluid flow regime and the convectional heat transfer mechanism. Even though empirical correlations can often provide accurate estimates of convective heat transfer, some specific conditions require different strategies. This can be due to the condition where the correlations have insufficient accuracy or the current boundary conditions required for applying the correlation are unknown. As mentioned before, the local heat transfer coefficient can be determined based on the temperature distribution within the fluid using equation 2.7. Convective heat transfer strongly depends on fluid flow behavior, as the flow patterns influence the temperature distribution.

Therefore, numerical fluid dynamics (CFD) is a method for solving the governing equations of fluid flow presented in this section. J. Anderson [32] addresses respective details for various studies due to the wide range and complexity of fluid dynamics in engineering applications. The governing equations for the fluid flow and heat transfer models are solved using numerical methods. In this case, the governing equations in the fluid domain are the Navier-Stokes equations which are conservation equations of mass, momentum, and energy [32]. The equations derived are presented in the following section.

2.2.2.1 Conservation of Mass

Considering a control volume of arbitrary shape and a finite size fixed in the fluid domain, the forms of the governing flow equation can be obtained from the specified element in the space called conservation form. The fluid moves through the fixed control volume across the control surface [32].

For a system,

Time rate of change of the mass of the coincident system	=	Time rate of change of the mass of the contents of the coincident control volume	+	Net rate of flow of mass through the control surface

The net mass flow rate through the control surface is stated in equation 2.18.

$$\int_{CS} \rho v \vec{n}\, dA = \sum \dot{m}_{out} - \sum \dot{m}_{in} \qquad \text{Equation 2.18}$$

Therefore, for the fixed, nondeforming control volume the conservation of mass is defined in equation 2.19.

$$\frac{\partial}{\partial t}\int_{CV} \rho\, dV + \int_{CS} \rho v \vec{n}\, dA = 0 \qquad \text{Equation 2.19}$$

Equation 2.19 is the partial differential equation form of the continuity equation. The first term of equation 2.19 describes the net flow of mass across its boundaries, known as the convective term. In steady conditions, the first term is connected to the time is zero. Therefore, equation 2.19 presents the unsteady, three-dimensional mass conservation.

2.2.2.2 Conservation of Momentum

The conservation of the extensive property momentum follows Newton's second law. Newton's second law can be employed to interpret the momentum equations as the change in momentum connected to the sum of forces for a particular oil particle. Therefore, the following statement is carried out for a system.

<table>
<tr><td>Time rate of change of the linear momentum of the system</td><td>=</td><td>Sum of external forces acting on the system</td></tr>
</table>

Surface forces like pressure, viscous forces, and body forces like gravity, centrifugal, Coriolis, and electromagnetic forces are frequently present on fluid-particle surfaces. The time rate of change of the linear momentum is expressed as the sum of two terms.

<table>
<tr><td>Time rate of change of the linear momentum of the system</td><td>=</td><td>Time rate of change of the linear momentum of the contents of the control volume</td><td>+</td><td>Net rate of flow of linear momentum through the control surface</td></tr>
</table>

This can be also expressed in terms of equation 2.20 [32].

$$\frac{\partial}{\partial t}\int_{cv} v\rho dV + \int_{cs} v\rho \vec{v}\vec{n} dA = \sum F \qquad \text{Equation 2.20}$$

The first term on the left side represents the time rate changes of linear momentum of the contents of the control volume, and the second term shows net rate of flow through the control surface. For the steady flow, the time rate change of the momentum is zero. The right term of equation 2.20 presents the acting forces on the control volume of flow.

2.2.2.3 Conservation of Energy

The first law of thermodynamics is proved that energy must be conserved. It states that a fluid particle's energy change rate equals its additional heat rate plus the work done on the fluid particle. For a system,

$$
\begin{array}{ccc}
\begin{array}{c}\text{Time rate of increase}\\ \text{of the total stored}\\ \text{energy of the system}\end{array}
& = &
\begin{array}{c}\text{Net time rate of}\\ \text{energy addition by}\\ \text{heat transfer into the}\\ \text{system}\end{array}
\quad + \quad
\begin{array}{c}\text{Net time rate of energy}\\ \text{addition by work transfer}\\ \text{into the system}\end{array}
\end{array}
$$

In symbolic form, this statement is given in equation 2.21.

$$
\frac{D}{Dt}\int_{sys} e\rho dV = \left(\sum \dot{Q}_{in} - \sum \dot{Q}_{out}\right)_{sys} + \left(\sum \dot{W}_{in} - \sum \dot{W}_{out}\right)_{sys}
\qquad \text{Equation 2.21}
$$

The total stored energy per unit mass for each particle in the system, e, is related to the internal energy per unit mass, $\hat{u}$, the kinetic energy per unit mass, $\frac{v^2}{2}$ and the potential energy per unit mass, gz by the equation 2.22 [32].

$$
e = gz + \hat{u} + \frac{v^2}{2}
\qquad \text{Equation 2.22}
$$

The net rate of heat transfer into the system is denoted with $\dot{Q}_{net,in}$, and the net rate of work transfer into the system is labeled $\dot{W}_{net,in}$.

The control volume for the first law of thermodynamics is expressed in equation 2.23.

$$
\frac{\partial}{\partial t}\int_{cv} e\rho dV + \int_{cs} e\,\rho v\vec{n}dA = \left(\dot{Q}_{net,in} + \dot{W}_{net,in}\right)_{cv}
\qquad \text{Equation 2.23}
$$

The heat transfer rate, $\dot{Q}$, represents all the ways in which energy is exchanged between the control volume contents and surroundings because of temperature difference. The net heat transfer rate can be zero in the process of adiabatic.

The work transfer rate, $\dot{W}$, called power is positive when the work is done. Power can be mostly transferred across the control surface by a moving shaft. A source of energy S_E defines per unit volume per unit time is used to add energy in the system.

2.2.2.4 Turbulence Modelling

In this subsection, the effects of the fluctuations associated with turbulence on the Navier-Stokes Equations are analysed. The velocity fluctuations give rise to additional stresses on the fluid, called Reynolds stresses.

The Reynolds-averaged Navier-Stokes (RANS) Equations can be obtained by applying statistical averaging to the turbulent flow. For engineering applications, knowing all the turbulent velocity and pressure variations in detail is usually unnecessary, and the averaged flow is typically sufficient. Therefore, the flow is considered more detailed and divided into average and fluctuation values.

The flow near the wall must be studied to comprehensively understand turbulence. The fluid contacting with the solid wall fulfils the no-slip condition, therefore at equation 2.24 applies at the solid wall.

$$u_w = v_w = w_w = 0 \qquad\qquad\qquad \text{Equation 2.24}$$

With u_w, v_w and w_w representing the velocities at the wall. The turbulent flow is predominantly anisotropic in the boundary layers. By increasing the distance to the wall, the influence of the wall decreases, and the turbulence becomes gradually isotropic. Velocities and non-dimensional distances are commonly employed to describe the general characteristics of the boundary layer, such as

$$y^+ = \frac{y.u_\tau}{v} \, , \, u^+ = \frac{u}{u_\tau} \qquad\qquad\qquad \text{Equation 2.25}$$

With the friction velocity

$$u_T = \sqrt{\frac{\tau_w}{\rho}} \text{ with } \tau_w = \mu \left(\frac{\partial u}{\partial y}\right)_w \qquad\qquad \text{Equation 2.26}$$

where the corresponding variable is at the wall and is represented by the subscript w, the term u refers to the velocity in the tangential direction, and the wall-normal direction is y.

The RANS modelling can be used to carry out the essential information of the flow structure. The field variables of the turbulent flow are initially divided using the Reynolds decomposition to determine the averaged information, as shown in equation 2.27.

$$\phi(x,\ t) = \overline{\phi\ (x,t)} + \phi'\ (x,\ t) \hspace{3cm} \text{Equation 2.27}$$

Generally, $\overline{\phi\ (x,t)}$ is the time-averaged part and $\phi'\ (x,\ t)$ refers to the fluctuation component. Figure 2.12 shows a typical trace of the axial velocity component at a given location in a time interval of t [33].

In this study, the decomposition of velocity is expressed likewise in figure 2.12 by

$$u = \bar{u} + u' \hspace{3cm} \text{Equation 2.28}$$

The mean value $\bar{u}$ represents an ordered fundamental current and is determined by integration over a sufficiently short time interval.

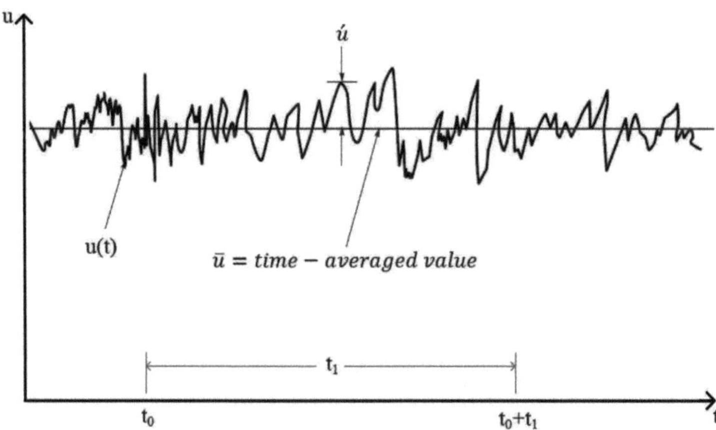

Figure 2.12: Description of a parameter for turbulent flow, the time-averaged $\bar{u}$, and fluctuating u' [33].

The conservation laws of mass and momentum are the principle that a flow obeys:

$$\frac{\partial \rho \overline{u_i}}{\partial x_i} = 0$$

Equation 2.29

$$\frac{\partial(\rho \overline{u_i})}{\partial t} + \frac{\partial(\rho \overline{u_i u_j})}{\partial x_j} = \frac{\partial \tau_{ij}}{\partial x_j} + \rho g_i$$

Equation 2.30

where ρ is the fluid density, u_i the velocity component in the direction x_i of the Cartesian coordinate, g_i the component of the gravitational acceleration in x_i direction, p the pressure and τ_{ij} the stress tensor.

Since an unknown term, called Reynolds stress tensor $\overline{u_i' u_j'}$ is present in the RANS equations, more information is needed to solve these equations.

For Newtonian fluids, the Reynolds stress tensor is denoted as τ_{ij}, is defined as in equation 2.31.

$$\tau_{ij} = -(p + \frac{2}{3}\mu \mathrm{div}\vec{u})\delta_{ij} + 2\mu D_{ij}\mu$$

Equation 2.31

Where

$$D_{ij} = \frac{1}{2}\left(\frac{\partial u_i}{\partial x_j} + \frac{\partial u_j}{\partial x_i}\right)$$

Equation 2.32

By introducing the dynamic viscosity, equations 2.29 to 2.32 are collectively known as the Navier-Stokes equations.

The Reynolds stress sensor refers to the effects of the fluctuation term on the averaged part. It can be carried out from two different methods: the Reynolds Stress Equation Model (RSM) and the Boussinesq approximation [34].

The RSM approach obtains satisfactory results when dealing with anisotropy, separation, or recirculation flows. However, solving the additional Equations is not preferable in most engineering applications as it leads to higher computational cost.

The other approach is called the eddy viscosity model, where the Boussinesq approximation is used to obtain the Reynolds stress tensor.

The Reynolds stress tensor is, then, approximated by the production of an eddy viscosity and the mean strain-rate tensor in equation 2.33.

$$-\rho \overline{u_i' u_j'} = \mu_t \left(\frac{\partial \overline{u_i}}{\partial x_j} + \frac{\partial \overline{u_j}}{\partial x_i} \right) - \frac{2}{3} \rho \, \partial_{ij} k \qquad \text{Equation 2.33}$$

where μ_t is the eddy viscosity and k the turbulent kinetic energy, defined as

$$k = \frac{1}{2} \overline{u_i' u_j'} \qquad \text{Equation 2.34}$$

The dynamic turbulent viscosity μ_t is a scalar, indicating that the turbulence is assumed isotropic, which can cause inaccurate results in anisotropic turbulent flows. Nevertheless, the eddy viscosity models are widely applied for engineering problems due to the lower computational cost by maintaining the accuracy of the results. Many different turbulence models have been developed to determine the dynamic turbulent viscosity μ_t.

The simplest models are the mixing length model proposed by Prandtl [35] and the Spalart-Allmaras model [36]. The two-equation models utilize two additional equations to reproduce the turbulent viscosity, yielding to accurate analysis of the turbulent flow modelling.

Various assumptions that yield to define these scales and the type and number of equations used for this purpose establish a classification scheme within the class of eddy-viscosity methods. Within the CFD – software CFX, applying $\text{SST} \, k - \omega$ turbulence model is described in Appendix A.2. The turbulence theory has been examined in several past studies [37-38].

2.3 Thermal Modelling of Power Transformers

Transformer load capacity is limited by the temperature of the winding, which must be kept under the maximum permissible temperature [39]. Excessive heat losses lead to faster insulation degradation and lower breakdown voltage.

Therefore, the temperature of the winding has a remarkable influence on the lifetime and affects the reliability of the transformers. This can be achieved using a high-performance cooling system in the transformer to maintain the operating temperature within the allowable limits. This section reviews the state-of-the-art thermal modelling of power transformers. The following sections will also highlight some of the attempts to study the thermal characteristics of transformers.

2.3.1 Thermal Performance

The thermal performance of power transformers is a critical aspect of their operation and reliability within electrical power systems. Power transformers are subjected to significant heat generation due to the inherent losses in their core and winding materials. Efficient dissipation of this heat is essential to maintain safe operating temperatures and prevent overheating, which can lead to failures or reduced lifespan. Effective thermal performance management involves the design and implementation of cooling systems. Advanced thermal modeling and monitoring techniques play a pivotal role in understanding and optimizing this performance.

By continually assessing and managing thermal aspects, power transformers can operate within their designed temperature limits, ensuring reliable electricity transmission and distribution while prolonging their operational lifespan. Thermal performance of power transformers can be determined using three different approaches: thermal modelling, measurement devices and heat run tests. Thermal modelling techniques can be divided into three major categories, including thermo-circuit analogy and network models.

Traditional methods of transformer winding thermal design use lumped parameter (semi-empirical) models such as 'thermal-electric analogy' [40-43] and 'network models' [44-46]. The third category, computational fluid dynamics (CFD), is based on the highly discretized finite volume method (FVM) or finite element method (FEM). Figure 2.13 shows a framework of the research themes covered by the literature review. Thermal modelling is beneficial in predicting the hot spot temperature and is necessary during the design stage or for new operational loading scenarios. The heat run test is based on the principle of calculating the total loss. It is categorized into two approaches, short-circuit for calculating copper loss and open-circuit for calculating iron loss.

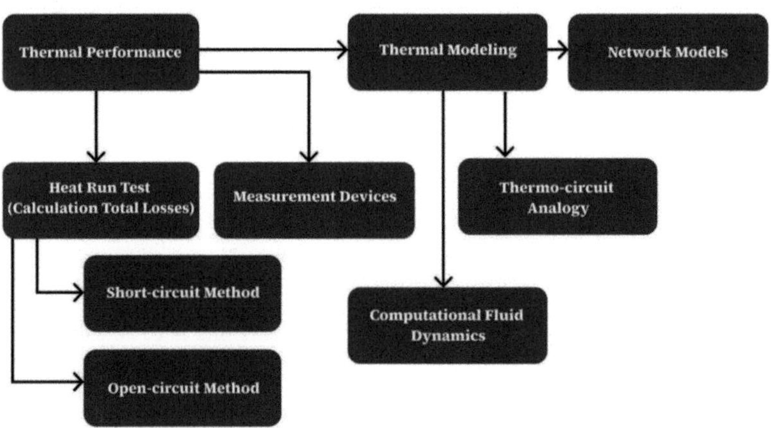

Figure 2.13: Framework of research themes covered by the literature review.

For large power transformers, heat run tests are limited because they can only assess the global temperatures, such as top oil temperature and average winding temperature. Therefore, the necessity to install optical fibers for the temperature measurement of the hot spot temperature is promoted [39].

2.3.2 Basics of Thermal-Hydraulic Network Model (THNM)

The design and operation of power transformers require a comprehensive research, which increases the need for advanced thermal modelling tools and direct temperature measurements [47-49]. In this regard, CIGRÉ Working Group A2.38, formed in 2008, has included state-of-the-art thermal modelling techniques. The study is concluded that CFD and THNM represent adequate techniques within thermal modelling [1].

Due to the importance of these methods for defining temperature distribution, the thermal-hydraulic network model is explained in this section. Detailed THNM modelling relies on basic conservation principles as follows:

a. Conservation of heat

b. Conservation of mass

c. Pressure equilibrium in a closed loop

Unlike CFD modelling, which is a distributed parameter approach as the complex system of partial differential equations, THNM describes them by a set of equations.

The detailed global THNM model of core type windings comprises two interdependent models; (1) the Hydraulic Network Model; (2) Thermal Network Model.

The basics of the THNM model are expressed through the following steps [46]:

- Calculation of driving force in a closed loop:

Simplest Single Oil Loop: The pressure drop appears due to oil flow throughout the channels during the oil circulation. The thermal driving force appears after changing density in the closed oil loop. In steady-state oil flow, the flow rate has values so that the equilibrium of produced pressure and total pressure drop exists. Figure 2.14 shows a simple single oil loop [46]. Oil is heated in the winding (AB), flows through the pipeline above the winding (BC), is cooled within the radiator (CD), and flows through the space between the output from the radiators and the winding (DA). In the area BC and DA, the heat exchange occurs and it is negligible compared to the heat exchanged in windings and radiators [46].

The thermal driving force (P_T) is defined in equation 2.35.

$$P_T = \oint \rho \vec{g} \, \vec{dl} = \oint \rho g \cos \phi \, dl \qquad \text{Equation 2.35}$$

where ϕ is the angle between velocity and gravity vector and $\vec{l}$ is path vector.

For representing the simplified form, the thermal driving force is calculated as

$$P_T = \rho_r g \beta \Delta \theta_{ol} \Delta H \qquad \text{Equation 2.36}$$

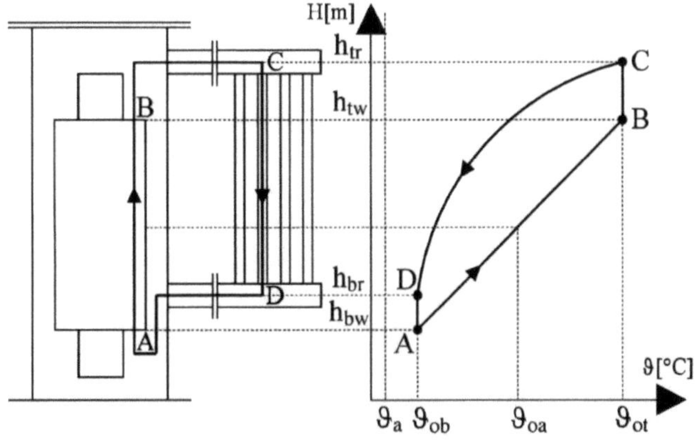

Figure 2.14: Change of oil temperature along the simplest oil loop [46].

ϑ_a: Ambient Temperature (°C), ϑ_{ot}: Top Oil Temperature (°C)

ϑ_{ob}: Bottom Oil Temperature (°C), ϑ_{oa}: Average Oil Temperature (°C)

where ρ_r is the oil density at referent temperature and $\Delta\theta_{ol}$ is the vertical temperature gradient between top and bottom of the winding.

ΔH is the height difference between the center of the winding and the center of the radiator. In the OD cooling condition, the pressure produced by pump is added to the thermal driving force. By ON cooling condition, only the thermal driving force attends in the oil closed loop.

- Calculation of thermal resistance:

Heat transfer can be expressed by simple expressions for thermal resistance. For the plane wall, the thermal resistance of heat conduction is defined in equation 2.37.

$$R_\lambda = \frac{1}{\lambda} \cdot \frac{\delta}{A}$$

Equation 2.37

where λ is the thermal conductivity of the wall and δ is the thickness of the wall. For the convection, the thermal resistance is calculated in equation 2.38.

$$R_{Conv.} = \frac{1}{h.S} \qquad\qquad\qquad \text{Equation 2.38}$$

where h is the heat transfer coefficient and S is the attached surface to oil.

2.3.2.1 Hydraulic Network Model

A network-based hydraulic model can determine the mass flow distribution and pressure loss within disc type power transformers. It is assumed that the oil temperature in all axial cooling channels on the entry oil domain is uniform along the flow direction. This indicates that heat transfer from the conductor to the oil in axial cooling channels is neglected [2]. In this light, it should be noted that the local and frictional pressure drops are calculated for the oil flows and updated for local oil temperatures.

The hydraulic network is non-linear since the hydraulic resistances depend on the flow and must be solved by specific numerical iterative procedures. An iterative process of solving the complex network of the transformer determines the oil flow through each of its oil channels.

The total pressure drop across each element is calculated within the network, depending on its flow. The flows should be distributed in the iterative procedure to achieve pressure equilibrium in each independent oil loop [1].

Hydraulic networks can be explained using an example of a disc type winding with washers, and one example of the winding segment is given in Figure 2.15. The hydraulic network, as shown, consists of eight nodes.

Four nodes are selected for entrance, and four nodes are located at the outlet. One node at each side is placed at the oil entrance and outlet. The three remaining nodes on each side represent the merging and branching points. The summation of static and dynamic pressures is associated with the nodes. Friction hydraulic resistances describe friction-related pressure drops. The resistances are illustrated in Figure 2.16. The analogy with electrical circuits and the current flows due to completed oil flows.

a. Axial cooling channels on the left side: R_{af1}, R_{af3}, R_{af5} and R_{af7}

b. Axial cooling channels on the right side: R_{af2}, R_{af4}, R_{af6} and R_{af8}

c. Radial cooling channels – R_{f1}, R_{f2}, R_{f3} and R_{f4}

Local pressure drops are described by hydraulic resistances such as:

- Branching ($R_{C,\,s}$ and $R_{C,\,St}$)

- Merging ($R_{C,\,s}$ and $R_{C,\,St}$)

- Corner (R_C)

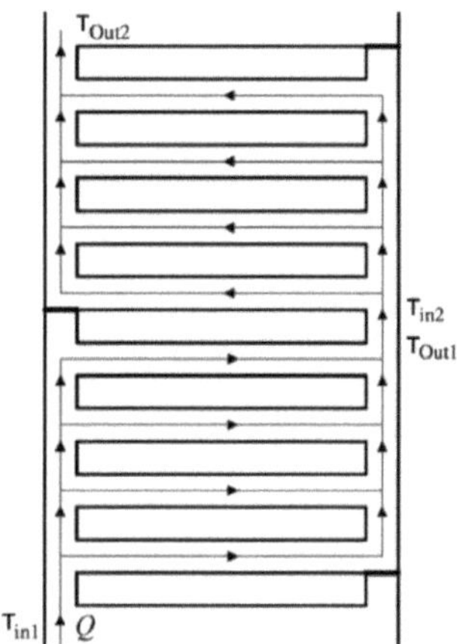

Figure 2.15: Illustration of a disc type winding with washers used for the arrangements of the nodes [46].

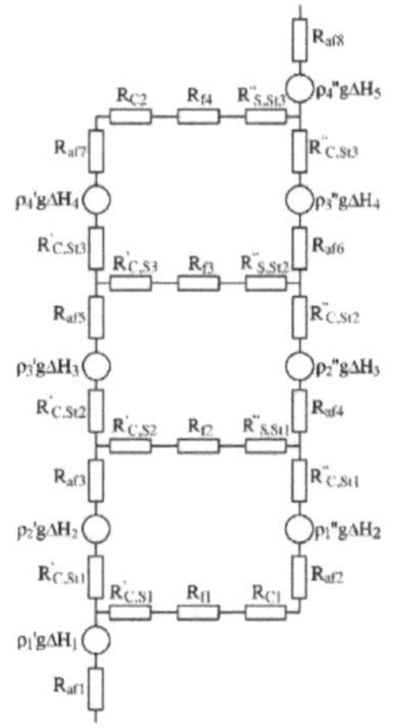

Figure 2.16: Hydraulic network of a pass in a disc type winding with washer [46].

T_{in1}: Temperature of oil entering the first pass (°C),

T_{out1}: Temperature of oil exiting the first pass (°C),

T_{in2}: Temperature of oil entering the second pass (°C),

T_{out2}: Temperature of oil exiting the second pass (°C),

Notably, the frictional and local pressure drops are computed for the oil flows and used to calculate the local oil temperatures. Since the hydraulic resistances depend on the flow, the hydraulic network cannot be linearly solved and requires some iterative numerical procedures.

The influence of oil flow on the heat transfer mechanism and exporting the heat energy from the winding discs should be considered. Therefore, a thermal network model is needed to couple the hydraulic network with the heat transfer mechanism.

2.3.2.2 Thermal Network Model

After introducing the hydraulic model, this section addresses the necessity of using thermal network modelling and discusses the details of a thermal network model. The heated oil through the conductors due to heat loss dissipations on the discs is achieved by coupling the thermal network model. This approach incorporates all heat transfer mechanisms into the hydraulic model. Published THNM investigations have employed different accuracy levels for each domain. The thermal network represents each conductor by one node, neglecting the difference in power losses and resistance to heat conduction inside the conductor.

Figure 2.17 shows the structure of the designed disc type winding, and Figure 2.18 shows the thermal network. The modelling considers the resistance between the insulation and the conductors, the heat convection between the oil and the solid domains, and the oil temperature change within the radial channels.

In the analogy of thermal network modelling, some simplifications are assumed. For example, the oil temperature in the axial cooling channels is the same along the flow direction. This means that heat transfer from the conductor to the oil in the axial cooling channels is not carried out by determining the flow and temperature distribution of the oil. According to [46], such a simplification is acceptable because it does not affect the local pressure drops or friction noticeably. A significant advantage of this simplification is that it enables solving non-linear equations. Thus, the temperature of the conductors can be established and calculated independently.

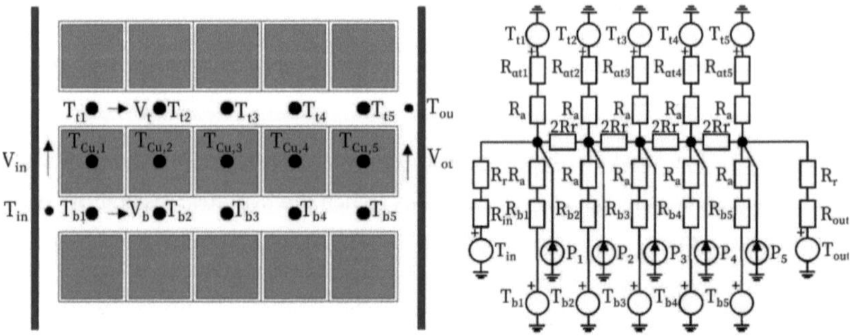

Figure 2.17: Selected disc with the nodes [46].   Figure 2.18: Thermal network modelling of a
                                                                                    disc [46].

After the initialization of oil properties, including the velocities and the temperatures, the algorithm substitutes the oil properties and calculates the new oil velocities using the hydraulic network. After providing the oil velocities, oil temperatures within channels are carried out using the thermal network. The combination of new temperatures and new velocities is used to update the oil properties for the next iteration. The calculation will continue until the relative changes in the oil velocities and the temperatures between two consecutive iterations decrease to the tolerance criteria. Finally, the wall temperatures and the temperatures of the conductors can be derived according to the oil bulk temperatures and the heat transfer mechanism.

Figure 2.19 summarizes the steps of solving network models using this algorithm. Following up on the steps during the calculation using this algorithm is more straightforward.

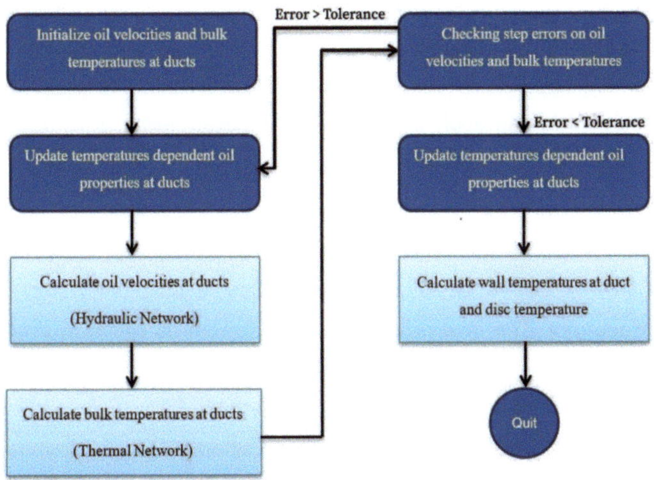

Figure 2.19: Algorithm for solving network models.

2.3.2.3 Literature Review of THNM

Oliver [44] has conducted one of the most comprehensive investigations and introduced a network model in the 1980s. His work presents the network model predicting the hot spot temperature and location. TEFLOW version 1 is developed to implement iterative solutions for the equations.

The study uses a calculation of a particular LV winding design as an example and shows that the hot spot temperature is located on the middle disc of the topmost pass of the winding. Past studies [50-51] also provided complete sets of mathematical equations for network modelling. For instance, the study [50] presents a detailed solving procedure in its appendix. Furthermore, [50-51] focus on only one winding, and the empirical equations are adopted from general fluid dynamics and heat transfer mechanisms.

Wu et al. [23] analyses parametric study the design parameters which have significant impacts on flow distribution, average winding temperature and hot spot temperature. Their study employs TEFLOW to evaluate the Nusselt equation in network modelling. A network model implementation is developed, and the influence of different Nusselt equations on the temperature distributions is assessed.

In another study [52], a whole transformer is modelled to create a comprehensive network model instead of the conventional single winding model. Moreover, in [53], the comparison between the THNM and 2D CFD modeling is conducted. It is figured out that the THNM can apply the pressure drop even faster by achieving reliable results for both approaches. On the other hand, the impact of phenomena like flow eddies cannot be captured by THNM.

Another notable work in this field is by Zhang and Li [54-55], which has described the setup of their model. Subsequently, it is applied to parametric studies in CFD investigations [56]. The goal is to describe the whole oil circuit to accommodate natural circulation. This achievement is established in [57], extending the previous work. Another study is contributed by Rahimpour et al.

[58] described the effect of other parameters, including the cooling system height. This work is comparable to that of Radakovic and Sorgic [46], which presented a self-contained description of a closed-loop THM with an alternative algorithm for converging the network model evaluation.

2.3.3 Computational Fluid Dynamics Method

The advent of powerful computing resources facilitates discretizing a physical problem on a complex geometry. The resulting equations can then be solved on powerful computers based on numerical algorithms capable of handling huge matrices arising from discretization.

CFD is based on the solution of the governing Navier-Stokes equations, which are derived from the conservation of mass, momentum, and energy for a fluid flow. Moreover, CFD simulations allow the characterization of the computational domain's pressure, velocity, and temperature fields, facilitating a complete understanding of the complex fluid phenomena within the transformer. However, the main drawback of CFD is that it is more computationally expensive and time-consuming than THNM and other analytical methods.

2.3.3.1 CFD Modelling Approach

The partial differential equations cannot be solved analytically, except for special cases. Hence, it is necessary to use a discretization method, which converts the set of differential equations to a system of empirical equations. In this regard, differential equations can be solved with a computational source, and a solution at different points can be obtained in the domains.

The numerical CFD calculations are listed in Figure 2.20. The discretizing process can start after defining the geometry. This is followed by defining the material properties and the boundary conditions. Consequently, the initial steps of the CFD procedure are providing the physical laws and the governing equations. Finally, post-processing is carried out after CFD calculations to visualize CFD results.

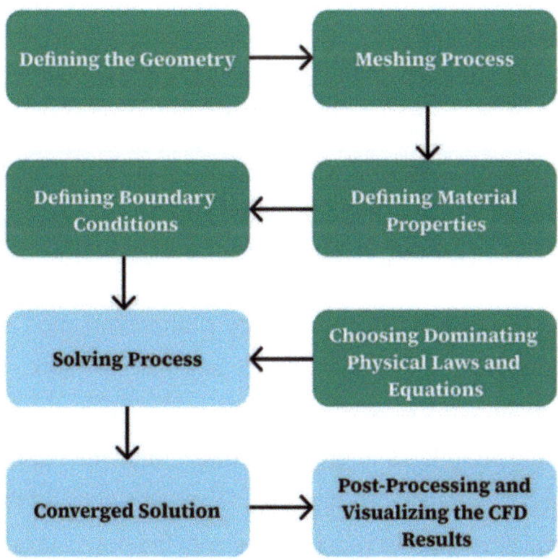

Figure 2.20: General steps during CFD solving procedure.

During this process, it is critical to determine the level of geometrical detail required to represent the problem adequately. For example, since an axisymmetric calculation is usually performed in power transformers, the presence of the radial spacers or the sticks can be neglected during calculations.

2.3.3.2 Literature Review Based on CFD Methods

CFD has been extensively applied to the thermal analyses of electrical motors and transformers. This effort replaces semi-empirical methods involving analytical formulas and constants which are derived from experimental results [59]. Over the last decade, thanks to the numerical simulations with CFD techniques, a better comprehension of the cooling mechanism of transformers has been applicable. Some recent trends in thermal modelling employ the coupled electromagnetic-thermal finite element models [60-61]. The literature on CFD calculations can be classified into 2D, and 3D CFD models, which is explained in the following section.

2.3.3.2.1 2D CFD Investigations

The initial steps of CFD modelling involve 2D simulations. As shown in [62], the 2D CFD investigation indicates the flow pattern results as a function of the Reynolds and Grashof numbers and the geometric parameters.

In [63], the transformer winding structure is examined using a 2D laminar flow model. The study considers density and viscosity as temperature dependent, and the loss at each conductor is determined based on the electric resistivity of the solid domain.

Shih [64] uses the geometry model, including half a core and two arrays of rectangular heating winding. An unstructured mesh is employed to figure out the temperature distributions. The study observes that the flow stagnation is located at the bottom of the winding, and the turbulent flow is visible at the top.

Additionally, a 2D CFD model is conducted by Kranenborg and Olsson et al. [65] to determine the significant effects of buoyancy and hot oil streak phenomena. A hot streak is formed from a streak of oil that flows along the disc surface and absorbs excess generated heat from the conductors. The hot streak can preserve temperature for a long duration due to the high Prandtl number of the oil. It is resulted that a fine discretization mesh is required to depict the hot streaks.

Studies like Weinläder et al. also perform a 2D CFD study to investigate the flow distribution within the horizontal channels [53-66]. The last two horizontal channels are determined to have a higher share of the oil, and it is illustrated that the eddies of the flow are visible at the entrance of the cooling channels. In [56], the dimensional analyses are considered in 2D cooled windings in both OD and ON cooling modes to identify the influences on the pressure drops and the hot spot factor.

2.3.3.2.2 3D CFD Investigations

Smolka [67] employs a coupled CFD-electromagnetic module with an optimization model. A dry-type transformer is considered, and the methodology applies to the oil-immersed transformer.

Torriano et al. [68] investigate the requirement of a CFD simulation in the thermal modelling of transformers studied. Moreover, an approach for mapping the results of a 2D simulation to a 3D simulation is carried out. A detailed CFD study on a single pass of the LV winding in an ON-cooling mode transformer is presented by Torriano and Chaaban [69].

The study is concluded that the effects of buoyancy by the ON-cooling mode cannot be neglected. Furthermore, the approximation of the solid domain, i.e., the winding discs, as homogenous copper blocks, obtains an acceptable accuracy during the calculation. CFD simulations also allow the study of hot oil streaks in the domain, as shown in Skillen et al. [70].

Moreover, different cooling modes are considered in examining the influence of the design parameters to enhance efficiency. As stated, a hot streak is only visible in a fine discretized domain. Hence, it is important to figure out the oil lines. Therefore, Figure 2.21 depicts the hot streaks through the winding channels. It presents the 2D CFD results investigated in [53]. As illustrated, oil hot streaks are observed in the right vertical channel. Due to the high Prandtl number of the oil, the hot streaks flow upwards along the vertical channel constantly.

Figure 2.22 shows the 3D CFD investigation shows the hot streak line in the 3D horizontal channels. The thermal boundary layers that emerge along the horizontal channel depict the transmission of the generated heat. The influence of hot streamlines on the flow pattern is investigated in [71].

The study observes the presence of hot streaks at the cooling channels, and the oil re-enters through the lower horizontal channel. This leads to a locally increased oil temperature; consequently, the temperature rises at the bottom region of the pass, and overheated conductors are noticeable.

Figure 2.23 presents the reversed direction of the oil flow through the vertical channel. In this condition, the outlet of the horizontal channel works as the inlet of the oil flow. The reverse oil direction in the cooling channel impairs the heat transfer mechanism and the oil flows only in reversed direction at the first channel above the washer. The magnified view of the conductors and cooling channels shows the overheated region due to the reversed flow direction of the oil. Rosas et al. [72] present the 2D and 3D CFD models for the solution of thermal and electromagnetic equations for transformer modelling.

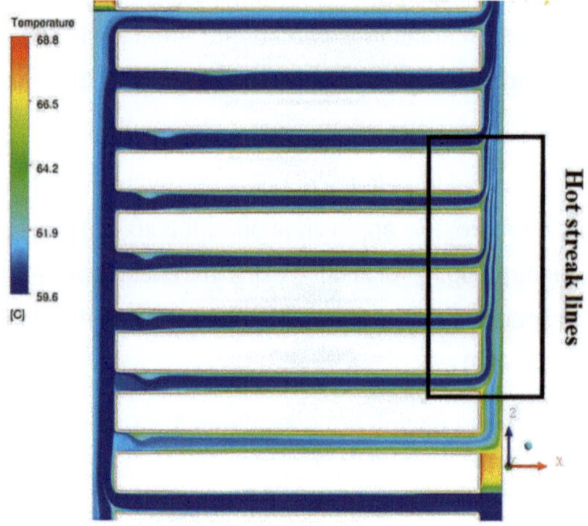

Figure 2.21: Illustration of hot oil streaks in 2D CFD model [53].

This aspect examines to improve the cooling process of oil-immersed power transformers using heat pipes. Ortiz [73] uses a coupled fluid flow and heat transfer modelling using a 3D CFD model for 3-phase transformers. The study determines the details of the active part and the tank geometry, including the fluid flow and electromagnetic field. Another study by Lefèvre et al. [74] examines a 3D FEM model using a magnetic scalar potential formulation investigated by combining a mixed analytical and numerical form of the electrical circuit equation.

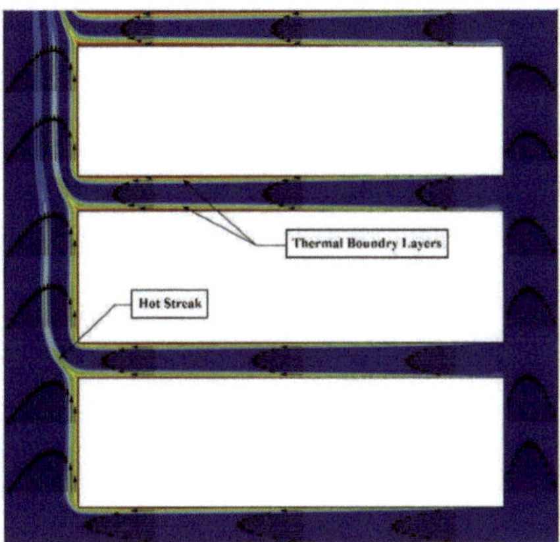

Figure 2.22: Illustration of hot oil streaks in 2D CFD model [69].

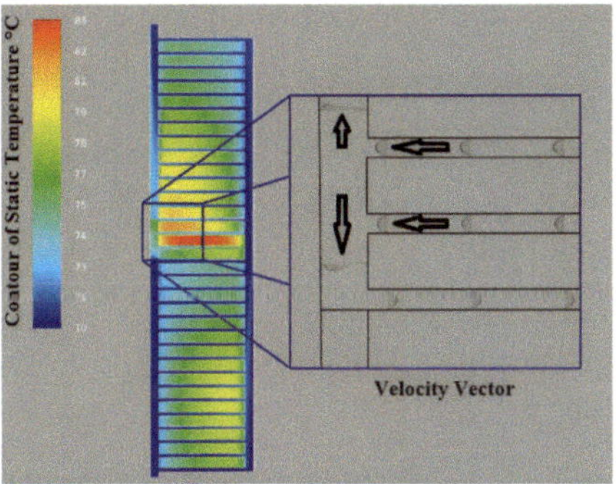

Figure 2.23: Temperature and oil velocity distributions within the channels, illustrating the overheated region and reversed direction of the oil flow [71].

Smolka et al. [75] examine a combination of fluid flow, heat transfer, and electromagnetic numerical analysis in three-phase dry-type transformers, and later, in [76], a 3D FVM is employed to model fluid flow in an encapsulated three phase dry-type transformer. It is coupled to an electric circuit model, and a detailed model for the electromagnetic field calculations is introduced, and in [77], numerical and experimental investigations of the temperature distribution inside the LV windings are carried out.

The evaluation of the thermal performances within the windings in 2D and 3D CFD modelling is studied by Tenbohlen et al. [78]. The measured temperature distributions at different flow rates are presented.

Both numerical 2D and 3D CFD approaches are compared with measured data to determine the influence of different modelling approaches. The result indicates that the location of the hot spot depends on the modelling approaches.

The 2D simplifications result in a partly high overestimation of the temperatures in the discs. In [78], at the flow rate of 9 kg/s, 10% difference between the temperature gradients is resulted between CFD 2D and 3D CFD approaches. On the other hand, at high flow rate, the numerical model overestimates the temperature of the conductors. In this light, 3D CFD approach is introduced as a robust method for simulating complex fluid phenomena, comparable with measured data. In this regard, Figure 2.24 presents the comparison between numerical CFD results and measured data. Here, the deviations of the temperature gradients in 2D at the flow rate of 18 kg/s with the same inlet oil temperature of 40 °C are even higher.

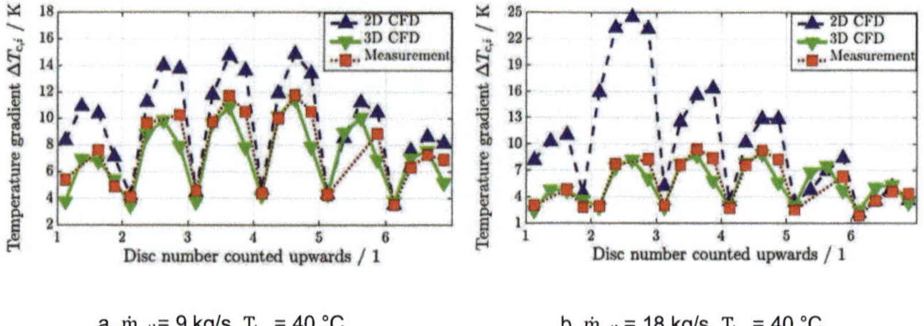

a. $\dot{m}_{oil}$ = 9 kg/s, T_{in} = 40 °C   b. $\dot{m}_{oil}$ = 18 kg/s, T_{in} = 40 °C

Figure 2.24: Comparison of the measured temperature gradients with numerically determined 2D and 3D CFD results at various oil flow rates and with specific losses per winding turn of loss = 360 W/turn [78].

The influences of eddy current losses on the thermal behavior of an OD cooled winding are examined in [79]. The study considers different geometrical parameters using different software to determine the hot spot temperature in the winding. It also examines the position of the hot spot and shows considerable impacts of non-uniform heat losses on the hot spot temperature. It is noted that the temperature distribution depends on heat losses and the arrangements of the washers.

The geometrical analyses show that decreasing the height of the horizontal channel, despite leading to an enhancement in the pressure drop, reduces the winding average temperature. Another study shows that despite the width of the vertical channels has a miner effect on the hot spot temperature, it reduces the pressure drop along the winding [80].

Biodegradable synthetic and natural ester oils are recognized as promising alternatives to mineral oils in power transformers [81]. Therefore, some recent investigations are focused on the thermal behavior of biodegradable oil liquids. Thermal analysis of an 8.5 MVA disc-type transformer cooled by biodegradable ester oil is studied by Stebel et al. [82]. Rapeseed ester oil is employed to determine the cooling efficiency of biodegradable oils compared with mineral oil in ON-cooling mode. Figures 2.25a and 2.25b indicate two different temperature distributions in LV and HV windings, using mineral and ester oil, respectively.

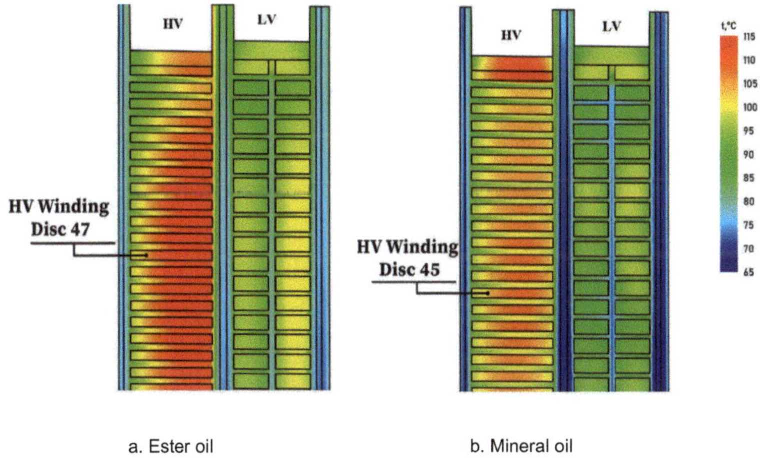

a. Ester oil   b. Mineral oil

Figure 2.25: Temperature distributions for two oil materials in the winding cross-section at the ambient temperature of 32 °C [82].

The ambient temperature is set at 32 °C to provide the hot spot temperatures in HV winding disc 47 by ester oil and HV winding disc 45 by mineral oil. The study is concluded that the hot spot temperature in LV and HV windings increases up to 9 K and 18 K by ester oil using ON-cooling mode. Moreover, it is illustrated that the winding temperatures are more uniform using mineral oil.

Due to the worse oil circulation by ester oil in ON-cooling mode, the hot spot temperature of the HV winding increased significantly in the ester oil [82].

Their other work [83] evaluate the velocity distributions inside the windings oil channels, and the local hot spot temperature by ester oil is determined. The results show that the average winding temperature is higher from 2 K to 9 K when ester oil is employed by ON cooling mode. In another study, a single-phase transformer is retro-filled with ester oil to test the thermal cooling capabilities of alternative transformer oil liquid [84].

The temperature test results are compared with the ON-cooled transformer, filled with mineral oil. The temperature rise test results for the two tested oil liquids are compared. It shows that the temperature of ester oil is lower than mineral oil only at the bottom region of the winding. Furthermore, the maximum temperature for ester oil is simultaneously higher than mineral oil.

2.3.3.3 Experimental Validation of CFD Models

Experimental tests have been performed following the numerical considerations to validate the numerical CFD results. Various devices have been employed to measure the oil flow velocity in practical methods, including hot wire anemometry (HWA) and laser-doppler velocimeter. Yamaguchi and Kumasaka et al. [85] have used a laser-doppler velocimeter to measure the oil velocity at the inlet channels of an ON-cooled winding model.

As the Doppler equipment is often more expensive and complex to set the system's coordinates, an alternative device like HWA can be employed due to the lower cost. HWA is also considered by Allen and Szpiro et al. [86] for measuring oil velocities inside the windings, compared with particle image velocimetry (PIV).

Daghrah et al. [87] determine the hot spot factor using the heat run test in an ON-cooled winding model. The experimental study depicts the dependency of the hot spot factor on the heat loss distribution and the inlet flow rate. Another study [88] conducts dimensionless analyses to simplify the physical problems of pressure drop and oil flow distribution. PIV systems are employed to measure the oil flow distribution, and the recorded pressure drop over the winding is implemented for the validation process. Most analyses of the experimental operating conditions have shown a superior influence of the Reynolds number than the Prandtl number.

Furthermore, conductors with the highest thermal resistances strongly depend on the maximum Reynolds numbers, while the Prandtl numbers have a lower effect on the respective results.

A recent study by Tenbohlen et al. [89] is performed an optical investigation of the scope of thermal performances. This study offered insights into the oil flow distribution resulting from specific boundary conditions. Tracing particles are added to the oil flow and illuminated by high-power LEDs to visualize the oil flow behavior. The winding model for the experimental investigation in this study is illustrated in Figure 2.26a.

The winding model is kept in a transparent oil container to allow photographs to be captured during operations. The particle velocities are recorded by taking pictures with a pre-defined exposure time. An example of a photo using particle tracks, including the magnified view of the particle tracks, is shown in Figure 2.26b.

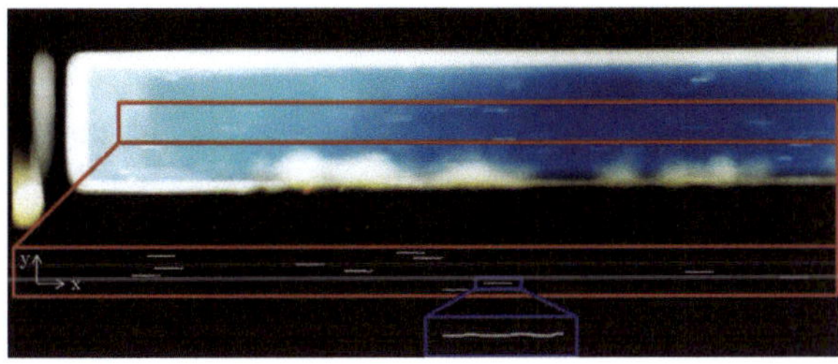

a. Experimental model in experimental investigation.

b. An example of a picture using particle tracks.

Figure 2.26: Details of the experimental setup for tracking particles [90].

This study illustrates that a more significant share of the oil flow is distributed to the last two channels of the investigated passes with increasing Reynolds numbers. The share of oil flows at the third and fourth channels is relatively independent of the operating conditions. In the last two channels, the operating conditions studied supplied 60-70% of the total oil flow, while the remaining oil flow is distributed to the following remaining channels. The higher flow rates indicate that the oil flow is distributed more homogeneously. Consequently, the temperature distribution is compared with the different cooling approaches. A technique for determining the mass flow rate of oil in an OD-cooled system is proposed by Pivrnec et al. [90].

The study uses two distinct disc types; in each case, oil flows into the transformer at the bottom and rises upwards within the winding while the hydraulic losses impacted the flow rate. Furthermore, calculating the oil flow velocity inside the channels simplified estimating the heat transfer rate in windings.

2.3.4 Thermal Monitoring and Online Measurements

On-site monitoring of transformer temperatures is a popular method for developing a strategy to maximize service life and the maintenance of power networks. The effectivity of the cooling systems directly affects the temperature within the winding and the lifetime of insulations. The loading capacity of transformers can be increased by early detection of failures based on an appropriate real-time online algorithm.

Therefore, online and offline monitoring systems have been developed to increase transformer operational efficiency and minimize the probability of an unexpected outage. The monitoring and on-site diagnostics of transformers have been considered for many years. A fiber-optic probe is the most common equipment for measuring local temperatures in the windings of the transformer. Thermocouple is also widely used to measure the temperature outside the winding block, like in the core and structural parts. Takala [91] presents some examples employing fiber-optic to measure hot spot temperature at different cooling designs.

Radakovic et al. [92] present several results for the distribution transformers with external cooling. The analysis of the measured results from the tested power transformers show that hot spot temperature rises over the top oil temperature. It is a function dependent on time as well as transformer load. In-service observations within the power transformers are performed at different ambient temperatures D. Susa and P. Picher [93]. The study concludes that using conventional top oil temperature models based on constant parameters can lead to underestimating hot spot temperature. This underestimation leads to an acceleration of ageing. Zhang et al. [94] perform an online temperature monitoring system using fiber Bragg grating sensors. The study prevents electromagnetic interferences using the sensors in an oil-immersed transformer, and realizes direct temperature measurement.

Consequently, an accurate estimation of transformer life loss is established by continuously integrating winding temperature and also measurement time. This estimation provides a reliable reference for transformer load regulation.

Feuchter et al. [95] install a microcontroller-based online monitoring system is installed on two coupling transformers at two local utilities. The calculation of hot spot temperature and life expectancy is performed using the algorithms, presented in standard IEC 60354. The load factor, the top oil temperature, the bottom oil temperature, and the ambient temperature can be measured and registered online. The study clarifies that the winding time constant should not be neglected during the calculation of the short-term behavior of the power transformers, especially under overload conditions.

Tenbohlen et al. [96] examine different aspects of the degradation processes in the insulation of the windings.

Followingly, in another study a comprehensive online monitoring system is proposed to diagnose the insulation health, and the installation of the sensors in [97]. The study applies the concept of overload capacity of the power transformers according to standard IEC 60354 in conducting thermal monitoring for part of their system. This thermal model is also used for load-dependent regulation of the cooling plant. It indicates that the need for cooling plant to transfer the losses to the environment. This is calculated as a function of the ambient temperature and the actual loading. While the summation of different measurable variables can be considered for online monitoring, it is rarely useful to consider the entire spectrum. Therefore, the sensor technology is adjusted to the specific requirements of a particular transformer, depending on the age and operational conditions.

Pudlo et al. [98] use the Standard IEC 60354 to calculate the maximum overload capacity of the transformer. It is integrated into one body model for an online monitoring system. Hence, the study argues that detailed monitoring information must assess maximal load and admissible overload operation time where it can be displayed in the diagrams of the power transmission system. This model has shown the ability to counteract the threatening failures in time and react with higher flexibility in critical systems and emergencies.

Radakovic et al. [99] investigate the hot spot temperature model provided by standard IEC 60354 by employing a monitoring system. Previous works establish the original thermal models for transformers by an ON cooled system [100-101]. The model determines the influence of non-linear thermal characteristics on the transient thermal processes instead of the exponential functions and the time constants.

Moreover, parameters of the model are selected from the measurements during a short circuit investigation.

The study notes that thermal characteristics of a transformer can be changed in a long-term transformer operation. Therefore, the initially determined thermal parameters can be changed from the values. Consequently, incorrect values during modelling can lead to an error in calculating the accurate temperatures.

Vilaithong et al. [102] examine the accuracy of three different models during the transient state of load current and ambient temperature. The study includes a model based on standard IEEE C57.91, the linear model and the one-body model. The results indicate that the estimated parameters are not constant values.

The different time intervals during observation and the various operating states of pumps and fans are noticeable. The study calculates the error between the calculated and measured maximum oil temperature to determine the applicability of the models. In the long-term investigation, the linear and one-Body models yield to more accurate results in the top oil temperature calculation than the standard IEEE models.

Furthermore, a model based on standard IEC 60354 is implemented, and the sufficiency of the optimized parameters in each model and the different datasets are investigated. The study asserts that the linear model presents the best thermal performance in the top oil temperature calculation for ONAN, ODAF, and OFAF transformers [103].

Castillo et al. [104] used the IEEE model C57.91-1995 to calculate transformer loss in a monitoring system. The main idea of their monitoring system is to increase and decrease the load by keeping the ageing factor and loss of life lower than a specific criterion.

Furthermore, Yun et al. [105] consider the determination of the overload condition of the transformer and its condition in an online monitoring system. The study presents the procedures and methods for transforming the actual load data into the overload characteristic curves, including estimating the overload basis for any ambient temperature and the overload of any magnitude and duration.

Djamali et al. [106] determine the loading capacity of an indoor distribution transformer by simplifying it to a linear relationship between the temperatures at the inlet and outlet of the transformation station and loading capacity.

The study suggested an empirical thermal model for the computation of the maximum oil temperature based on the calculation of the tank temperature to determine the loading capacity of the indoor distribution transformers. Figure 2.27 shows the ambient temperature, station temperature, top oil temperature and the calculation of the top oil temperature with the measured results for three months, while Figure 2.28 illustrates the monthly loading capabilities, used to create the model parameters [107].

This method is advantageous as the proposed thermal model can be used for already-installed transformers without removing them from service and on transformers without any temperature pocket.

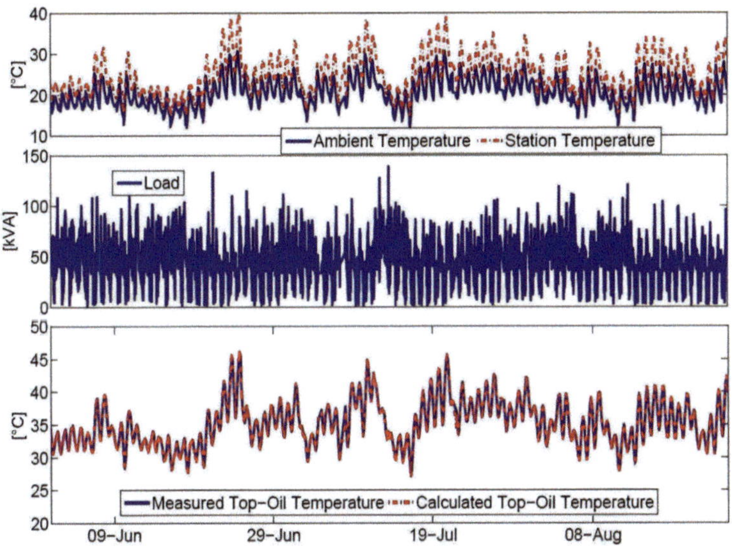

Figure 2.27: Measured ambient and station temperatures, measured load, measured and calculated top oil temperature using the proposed model [108].

Increasing the ambient temperature and station temperature can lead to a noticeable reduction in the loading capabilities. Furthermore, priori metrics are offered to select the best training data for top oil temperature models of the power transformer.

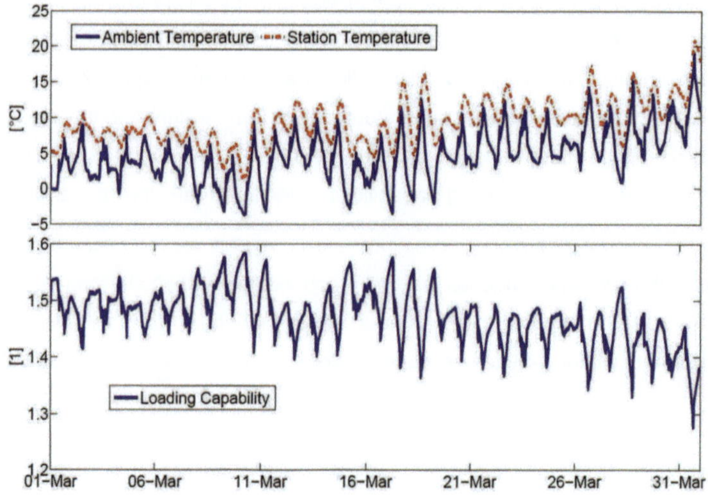

Figure 2.28: An example of calculating ambient and station temperatures and loading capability [108].

The method of investigation is applied to ONAN and ONAF and three top oil temperature models, including standards IEEE, IEC 60076-7 and linear model. It can be concluded that the best training data to train the model is achievable, and the distribution of standardized training error is closer to a normal distribution with zero mean and unit variance [107].

Several key conclusions from the literature review are listed below:

- Better thermal performance refers to the lower top oil temperature, average winding, and hot spot temperatures. The traditional assessment of thermal performance relies on the heat run test. However, only the global temperatures can be measured during a heat run test, and the hot spot factor roughly estimates the hot spot temperature. This highlights the need to measure the hot spot using optic fibers directly. Numerical modelling approaches are employed to predict the location of the hot spot temperature and assist the optic-fiber installation.

- The thermal-circuit analogy approaches provide faster results than network modelling but cannot anticipate the specifics of oil flow and temperature distributions. It leads to the approximation of the oil cooling system into several remarkable parameters.

- CFD simulations require significantly more computational effort than network modelling. Due to the better representations of details, CFD approaches can reveal some remarkable phenomena like hot streaks and flow eddies, which cannot be represented in network modelling.

- The 3D CFD approaches provide more accurate results than the 2D CFD investigations. Therefore, 3D CFD studies will be conducted for a detailed view of the thermal behavior of the fluid flows.

- Numerical analysis and network modelling necessitate experimental validation procedure. Key measurement devices for experimental investigations in the thermal behavior of power transformers include optical fibers, hot wire anemometry (HWA), Laser-Doppler velocimetry, and oil pressure sensors.

3. Development of CFD Numerical Winding Models

Although experimental investigations provide insights into the thermal performances of power transformers, it is necessary to provide numerical results to determine the effects of physical parameters. Apart from the high cost of constructing the experimental setups, experimental studies have limitations, such as the time required for an alternative model for most case studies. Obtaining different models and boundary conditions is essential to determine an optimized thermal model. Numerical investigation is a proper method for introducing the effect of geometrical or operational parameters. Hence, this chapter presents the 3D CFD force convection cooling modes. CFD method is used during numerical analysis to carry out the thermal behavior of power transformers. This method has shown high accuracy in depicting the results and corresponding with the experimental data. This chapter outlines the procedures for generating numerical 3D CFD models and discretizing geometries prior to initiating the computational process.

3.1 Determination of Sufficient Discretized Domains

3.1.1 Model A: Disc Type Winding Model with 4 Conductors per Disc

Two different winding designs are used during numerical investigation. Model one presents a disc type HV winding model with 20 discs. The front view of the passes, used during the investigation, is shown in Figure 3.1. Each pass is separated with washers (in pink) in the vertical channels, providing three different passes. It is observed that the oil flows upwards due to the convection mechanisms and divides into six horizontal channels. After flowing through the horizontal channels, the oil rises to the next upper pass. Therefore, the outlet of the first pass is connected to the inlet of the second pass. This structure is used for the third pass to provide a zig-zag disc type winding model.

Three different passes provide 18 discs; the last two discs are used to obtain an oil path to the outside of the model. A layer of insulation strips with the thickness of 3 mm (in light blue) is provided between two adjacent conductors. The material is selected due to the local ambient conditions. Due to the well corrosive resistance and low thermal conductivity, Murinyl, a registered trademark of Murtfeldt Plastics based on Polyvinylidenefluoride (PVDF) is chosen for insulation strips.

It has a mechanical stability up to the temperature 160 °C, suitable material for experimental approaches. It can be noted that the thickness of 3 mm for insulation stripes might be higher than the criteria of typical values for HV paper insulation. However, taking the range of thermal conductivity of oil impregnated insulation materials into account, it is assumed that the chosen design remains representative with regard to the thermal coupling of adjacent conductors inside a disc.

The location of the washer defines the direction of the oil flow, and the direction of fluid flow is illustrated in Figure 3.1, and the discs are numbered upwards. To ensure grid independence, the domain employs three distinct discretization levels. Consequently, fine mesh is chosen for computational purposes within the design.

Figure 3.2 shows the dimensions of the conductors and cooling channels with the insulation thickness in the mesh sensitivity analysis. The numerical model is exactly structured with the dimensions of the designed experimental setup. Figure 3.3 illustrates the position of the references at each discretized model, dividing the domains into fine mesh size, middle mesh size and coarse mesh size.

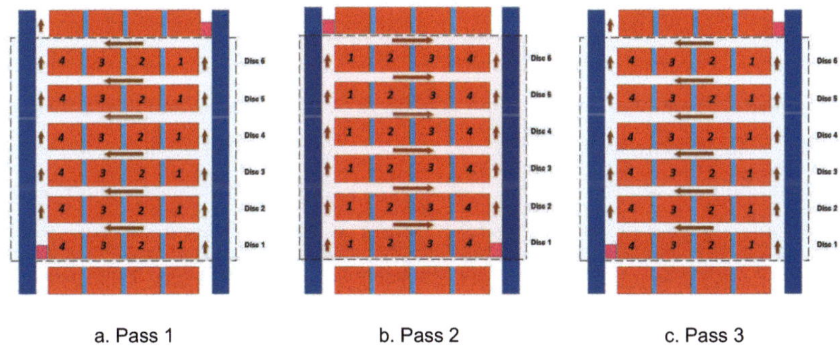

a. Pass 1   b. Pass 2   c. Pass 3

Figure 3.1: Front view of the investigated passes of the winding model A, with the direction of fluid flow and disc numbers.

Next to Figure 3.3, Table 3.1 gives the exact number of nodes at each position so that the number of elements at each domain is countable. During numerical investigations, this study only contributes the 3D CFD investigations, providing more details of the fluid flow behaviour.

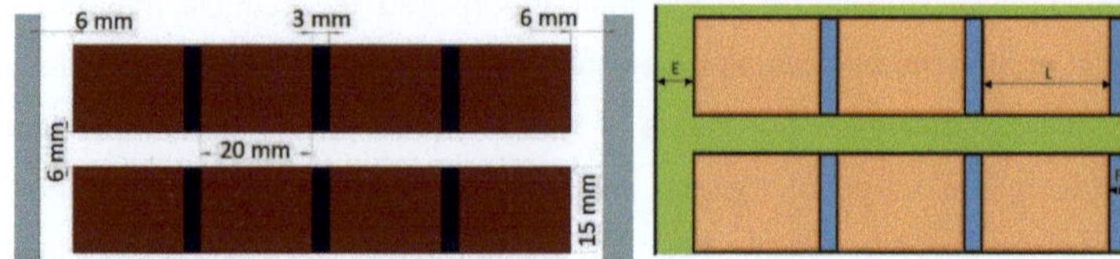

Figure 3.2: Overview of the dimensions.

Figure 3.3: Position of the references at each discretized model.

Table 3.1: Number of nodes at three different levels of discretization in model A.

Marked parameters	E	F	H	L
Number of nodes by discretization scheme in coarse size	18	8	34	40
Number of nodes by discretization scheme in middle size	36	16	68	80
Number of nodes by discretization scheme in fine size	54	24	102	120

Since the HV winding models are investigated in a broad range of operational conditions, a specific discretized model with the exact number of elements is needed. In addition, some phenomena of the fluid flows like hot streaks and flow eddies can only be illustrated with fine mesh and 3D CFD structure, this section gives a deep insight about the studied models. The vertical and horizontal cooling channels have the same number of nodes in each discretized domain. Figure 3.4 shows three different discretization levels at a common position of the winding model. The sub-layers close to the interfaces have extra nodes near to each other. It leads to have adequate nodes near the walls to model conjugated heat transfer.

The result of a CFD simulation improve when a mesh refinement is carried out because the values of the variables are calculated using points that are closer to each other. This means less error occurs when interpolation is carried out between adjacent cells or nodes that are closer to each other than when they are further apart. The reference numbers of the domains are also given in Figure 3.4, where position no. 1 refers to the conductor near the vertical oil channel, position no. 2 refers to the insulation, and position no. 3 refers to the horizontal oil channel. The boundary layer discretization is chosen with an initial cell height of $h_{init} = 10\mu m$.

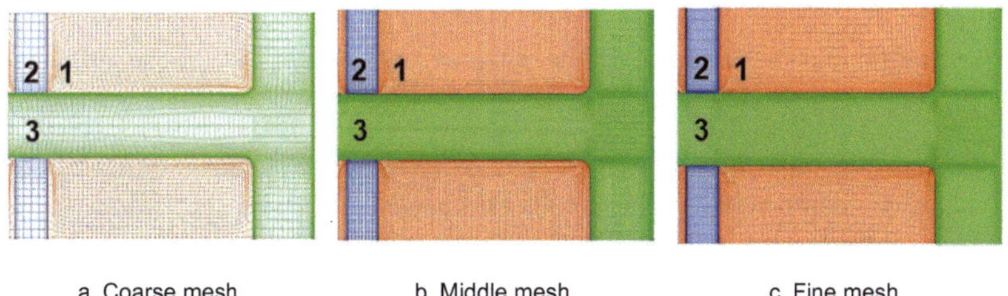

a. Coarse mesh   b. Middle mesh   c. Fine mesh

Figure 3.4: Discretization details at three different domains, no. 1 refers to the conductor, no. 2 refers to the insulation layer, no. 3 refers to horizontal and vertical channels, constant initial cell height $h_{init} = 10\mu m$.

Although 2D investigations require a significantly lower number of nodes and computation sources for CFD calculation, previous experiences in the field of thermal modeling of power transformers show that 2D CFD calculations gives higher deviations in calculation of temperature gradients with measurements.

Due to the rapid developments in the computational sources, by using geometry partitions and the applications of parallel calculations within the domains, the CFD results within the 3D designed model are achieved.

Model A is developed to aid in the experimental design of the winding model, providing valuable insights for the construction of Model B, which closely simulates the characteristics of a real winding of a power transformer. Detailed information about Model B will be presented in the following section.

3.1.2 Model B: Disc Type Winding Model with 8 Conductors per Disc

After introducing the discretization modelling in design called model A in Section 3.1.1, this section presents the disc type winding design with 8 conductors per disc, called model B. This winding model is also equipped with washers to provide a zig-zag pattern, and all conductors are covered with insulation material to study the effect of thermal resistance on insulation paper.

While model A includes only insulation material between adjacent conductors in a disc, mirroring the experimental model, model B features fully covered conductors and a greater number of conductors per disc to replicate the intricacies of a real transformer winding. The properties of solid materials are explained in Appendix A.1.

Figure 3.5 shows model B with 8 conductors per disc and two considered passes. The segment shows a detailed view of the inner and outer vertical channels. In this light, the position of the washers defines the number of cooling channels within each pass. Figure 3.6 provides the dimensions of the conductors and cooling channels to calculate the gap between nodes by the discretization process. Furthermore, Figure 3.7 presents the front view of the winding, including the entire parameters in grid generations.

The mesh design employs fine discretization for the investigations. The primary distinction between winding model designs A and B lies in the insulation layer and the number of conductors per disc. Model A, aligning with the experimental setup, consists solely of an insulation layer between the conductors. In contrast, model B covers the conductors entirely, necessitating additional calculations of the thermal resistance at all conductor interfaces (solid to solid and solid to fluid).

In Model B, the insulation paper between two adjacent conductors measures 1.2 mm in thickness, whereas the insulation paper between the oil and conductors is 0.6 mm thick. In this light, the researcher chose Model B to demonstrate the influences of additional vertical cooling channel in the winding designs.

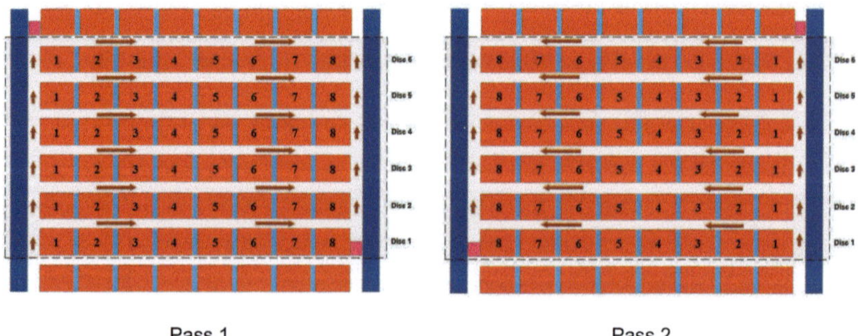

Pass 1.   Pass 2.

Figure 3.5: Front view of the investigated passes of the winding model B, with the direction of fluid flow and disc numbers.

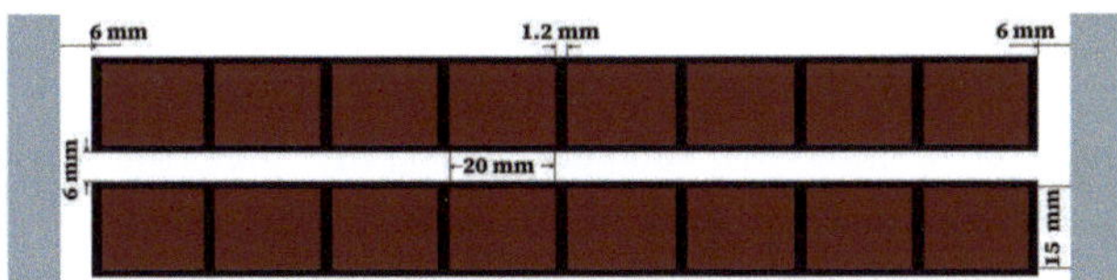

Figure 3.6: Dimensions of the disc type winding model.

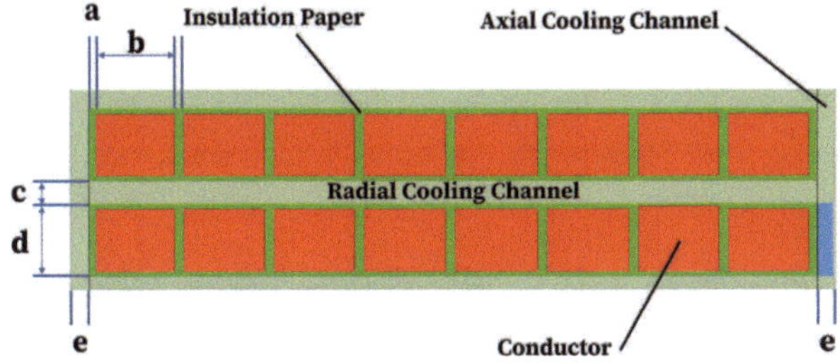

Figure 3.7: Position of the parameters in grid generation.

Table 3.2 gives the exact number of nodes at each position so that the number of elements at each domain is countable.

Table 3.2: Number of nodes at three different discretization levels in model B.

Parameter	Coarse	Medium	Fine
a) Thickness of the insulation paper close to the winding	4	6	8
b) Length of the conductor within the disc	20	30	40
c) Height of the horizontal channel between discs	20	30	40
d) Height of the conductor within the disc	20	30	40
e) Width of the conductor within the disc	16	24	32

3.2 Numerical Winding Model in Distribution Transformer (Model C)

After explaining the first two HV winding models (model A and B), this section is focused on the real distribution transformer design, called in this work, Model C. More details about the winding design and transformer capacity are provided on the distribution transformer's nameplate, which is included in Chapter 5, Section 5.1.1.

The numerical model is created based on the testing of a real distributed transformer. The HV winding is a disc type winding divided into five stacks with spacers. Each stack consists of 20 discs, with each disc having five turns. The conductors in HV winding have a thickness of 3 mm with 0.3 mm thick insulation paper.

The maximum radius of HV winding is 16 cm. Two layers are used in LV winding, where each layer has 19 segments with a width of 7 mm and a height of 21 mm, while the insulation paper around the LV conductors has a thickness of 2 mm. Figure 3.8 illustrates the central cross-section of the distribution transformer, while Figure 3.9 provides a top view showcasing the active components, including the limb and windings.

Figure 3.10 provides the magnified view of the LV and HV windings with the vertical oil channels. Lastly, the selected segment for the numerical investigation is illustrated in the red box in Figure 3.11. This segment is modelled due to the symmetrical thermal behavior of the distribution transformer design.

The model is created in the Solidworks 3D-CAD program, and grid generation is performed in ICEM Ansys. Two different methods are employed to generate mesh within the various domains. ANSYS Meshing software meshes the fluid region at the upper and lower parts of the transformer core using tetrahedral and prism elements. Prismatic boundary layer elements are used for all walls to achieve high quality elements near the walls.

The ICEM software generates mesh for the LV and HV windings, insulation papers and spacers. The same number of nodes connected the elements at the entire solid-to-solid interfaces (winding to papers and spacers). Based on the experience of grid generation of the disc type winding model in Section 3.1, the set characteristics of the selected meshes with the materials of each domain are presented in Table 3.3. Overviews of the generated elements for both HV and LV windings with the insulation papers are illustrated in Figure 3.12 and Figure 3.13, respectively.

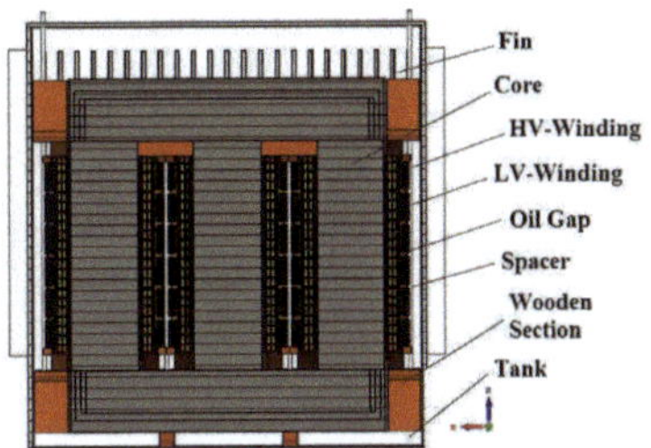

Figure 3.8: Middle cross-section of the distribution transformer.

Figure 3.9: Top view of the tank, including fins and windings.

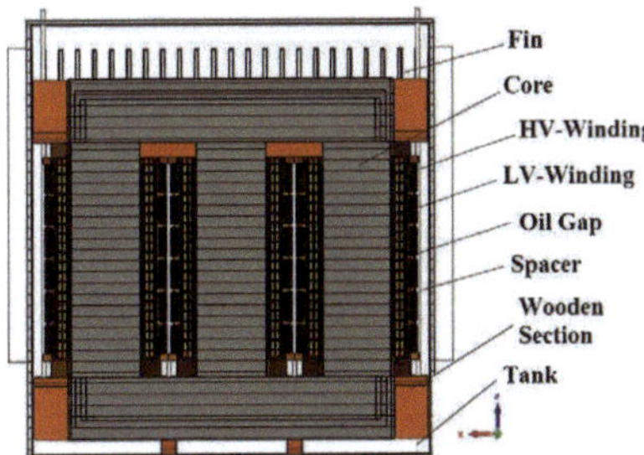

Figure 3.10: Magnified view of the windings with oil channels.

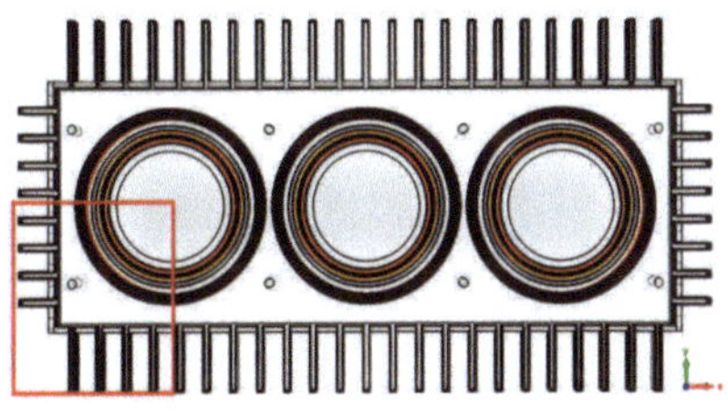

Figure 3.11: Top view of the selected segment for numerical 3D CFD investigation.

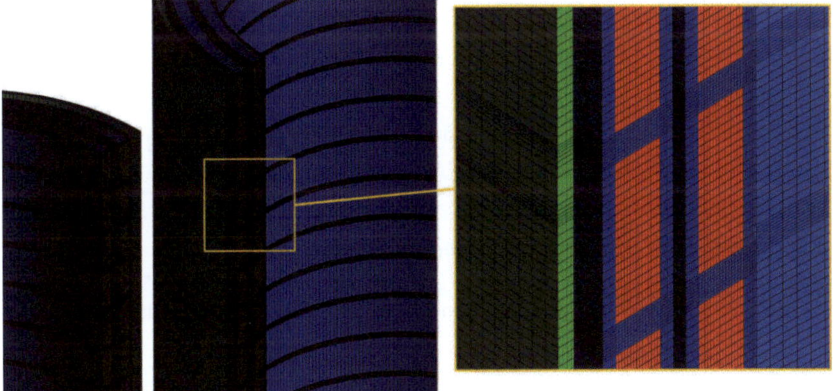

Figure 3.12: The generated mesh in ICEM software for LV winding with the insulation paper.

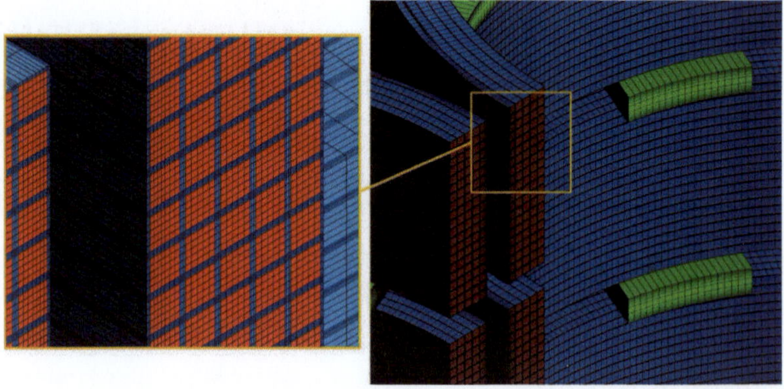

Figure 3.13: The generated mesh in ICEM software for HV winding with the insulation paper.

Table 3.3: Number of elements and materials in the numerical modelling of the distribution transformer.

Domain	Material	Number of Elements	Domain	Material	Number of Elements
Core	Steel	1,896,546	Transformer Fluid	Mineral Oil	39,761,549
HV Winding	Copper	4,568,400	LV Winding	Copper	2,534,220
LV Paper	PVDF	4,943,952	HV Paper	PVDF	8,527,680
Inner Insulation	PVDF	846,846	HV Spacers	Wood	461,700
Total number of elements in the numerical winding model			65,532,189		

3.3 CFD Model Setup and Solver Configuration

The set-up is imported into the selected CFD software product after appropriately discretizing the geometries described in the previous sections. All CFD-based investigations for this work are conducted with the proprietary software ANSYS CFX based on the FEM method.

ANSYS CFX is updated several times during the investigations, starting from version V.18 and going to V.22, with no noteworthy differences resulting from the software updates. In ANSYS CFX, the connected work steps related to a CFD calculation are handled by partitioning on clusters using complex computational sources.

The pre-processing of the modelling after grid generation is performed in CFX-Pre, followed by the calculation monitoring in CFX-Solver Manager. Consequently, the post-processing and visualization processes are handled in CFX-Post. Furthermore, the programming language Perl is supported within ANSYS CFX scripts, allowing additional improvements in script creation. The outline script-based approaches have significant advantages regarding large grid generations which is used in this work.

Concerning the model setup handled during pre-processing of the winding models, a comprehensive investigation is performed in [108] to determine the differences between CFD software, including Open FOAM and COMSOL Multiphysics. The study shows that ANSYS CFX, based on Finite Volume Method (FVM), has higher accuracy in 3D CFD calculations in modelling the thermal behavior of power transformers [108].

First, the model setup implements the oil temperature-dependent fluid properties used in the experimental investigations. Regarding the initial boundary conditions of the fluid domain, definitions are set for the oil inlet and outlet surfaces. During the forced convection analysis, a velocity inlet featuring a constant velocity over the entire inlet surface and a pressure outlet with a relative pressure of 0 Pa is selected. The chosen velocity corresponds to the investigated mass flow rate at the investigated operational condition. Buoyancy is represented with a reference density equaling the oil density at inlet conditions.

Furthermore, second-order formulations are set to increase the accuracy of the advection schemes. During the natural convection analysis, the Buoyancy is contributed using the Boussinesq approximation, and thermal force due to the density change by temperature is used. The boundary condition at the outlet is specified as a pressure outlet. The outer walls of the 3D CFD winding model are treated as adiabatic layers with a no-slip condition, while the outer walls of the distribution transformer are modeled with a specific heat transfer coefficient along the vertical walls of the fins to account for convective heat transfer through the fins.

Interpolation schemes consist of second-order upwinding for the convective terms and the second-order central difference for the diffusion terms.

The pressure velocity coupling is achieved using the SIMPLE method. Initial guesses are used for initializing the numerical model. Under relaxations are set adequately low in order to assist convergence. A transient analysis is performed, with a time discretization of 0.001 s (time step) and 20 iterations for each time step. Due to the various operating conditions, such as different inlet flow rates and inlet temperatures, the numerical CFD investigations are performed with laminar and turbulent flow patterns. The topic of turbulence modelling is already introduced in Section 2.2.2.4.

The $SST\,k-\omega$ turbulence model is chosen from the turbulence models focusing on the RANS-based approaches tested within the RANS-based models available within ANSYS CFX, details about the SST turbulence model are presented in Appendix A.2.

Several important factors support the selection of this turbulent model. First, it shows stability throughout all investigated boundary conditions in this contribution. Moreover, ANSYS CFX obtains additional extensions for the $SST\,k-\omega$ model which improves the suitability of transitioning flows.

Finally, it agrees well with experimental results within the investigated operating conditions presented in Chapter 4. For the convergence procedure, monitoring points are set at the conductor temperatures, the oil outlet and inlet temperatures, and the inlet pressure. While monitoring points are placed along the conductors, the process of solving energy and momentum equations is carefully executed and supervised. In this study, the convergence criteria and the root mean square (RMS) residual levels of 10^{-6} are selected during the solving iteration.

4. Investigation of Parameters Affecting Thermal Behavior of Oil-Directed Winding

This chapter analyses the significant operational and geometrical parameters to predict their influences on the thermal behavior inside an oil-directed cooled winding model. This chapter presents 3D numerical CFD results to determine the hot spot temperature and its location by applying various boundary conditions validated with measured temperatures from an experimental setup. The experimental setup consists of an oil feeding unit and a winding model comprised of 20 discs, equipped with fluid guides (washers) to obtain a zig-zag cooling mode.

Concerning the variations in the thermal characteristics of oil, the oil properties are changed by increasing the inlet oil temperature. Within this context, the geometrical parameters can be divided into horizontal cooling channel height, vertical cooling channel width and number of discs per pass. On the other hand, the primary operating parameters include inlet flow rate, oil temperature and heat loss distributions. Concerning the topic, the influence of parameters in operational and geometrical designs are explained in this chapter. Additionally, it presents an exploration of the thermal behaviour of different oil liquids, such as biodegradable natural ester oil. Subsequently, the following section introduces the fundamental designs of the winding geometries for both Model A and Model B, each discussed separately.

4.1 Basic Design of the Winding Geometries in Model A

The design of disc type windings shows a strong symmetry in a circumferential direction [109]. Figure 4.1 illustrates a magnified three-dimensional view of a segment with two main investigated passes to display these design characteristics. The entire model consists of three passes. A pass is defined as the number of discs between two successive washers. The first is the pre-conditioning pass, whose task is to obtain stabilized fluid flow for the investigations. A post-conditioning pass can be found at the top of the model, providing steady flow conditions at the outlet part during measurements. Similar to the experimental setup, each pass has six discs and four turns, and the experimental winding model is designed with 20 discs. Additionally, the magnified view includes the disc numberings used to present results. The numbering of the conductors within each disc is illustrated in Figure 4.1.

The preliminary design of the experimental winding should correspond to typical parameters used in power transformer design. Hence, transformer designers should choose the appropriate values to achieve this goal.

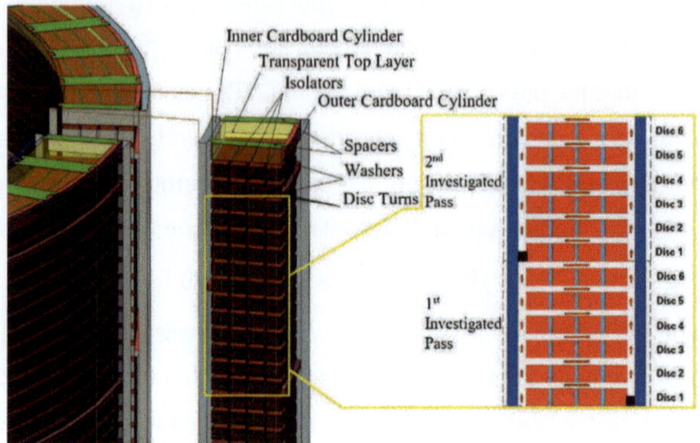

Figure 4.1: Basic winding design and alignment of spacers by creating symmetrical sections with periodic cooling channel system and numbering of the discs.

The rated power and voltage already establish a range for some features, even though the exact specifications of a given transformer winding mostly depends on demands, specified by the customer.

In this study, an inner winding diameter of 1.5 m is chosen to approximate an OD cooled HV-Winding range of a 200 to 400 MVA power transformer. The spacer interval in the circumferential direction is fixed at 8° following this diameter, reflecting the partitions of 45 segments. Figure 4.2a presents a top view of the winding model to describe the relationship between the symmetrical section and the whole winding model. The experimental winding model and numerical simulations are structured based on the symmetrical part of the transformer winding shown in Figure 4.2b before modification. Figure 4.2c presents the symmetric section, the modelled winding design, and the geometry dimensions. As seen in Figures 4.2b and 4.2c, the spacer geometry has been modified into a trapezoid form, and the shape of the spacers is kept as straightforward as possible due to the significant machining expenses associated with the materials in the experimental model. To keep the efforts connected to the production and assembly process of the winding model, while also enabling a proper investigation of the radial temperature distribution, four conductors in radial direction are chosen.

Besides, the surfaces between the oil paths and conductors are kept nearly identical to that of the real winding design. The width of the spacer is set at 3 mm, and the spacer height, which determines the height of the resulting horizontal cooling channels, is set at 6 mm. Additionally, the height and width of conductors are adjusted to 15 mm and 20 mm, respectively.

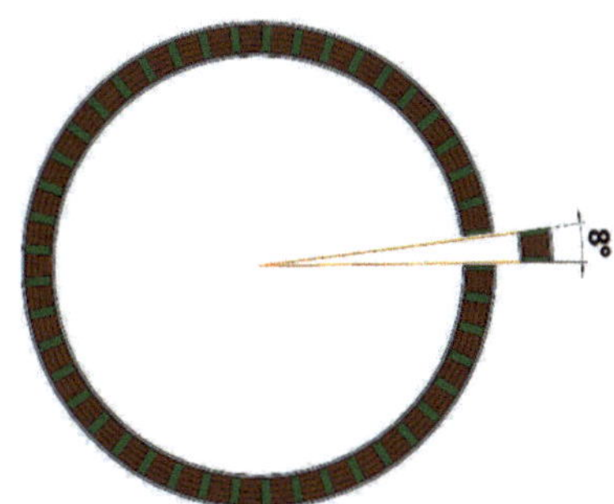

a. Top view of the selected segment, relationship between symmetrical section and whole windings.

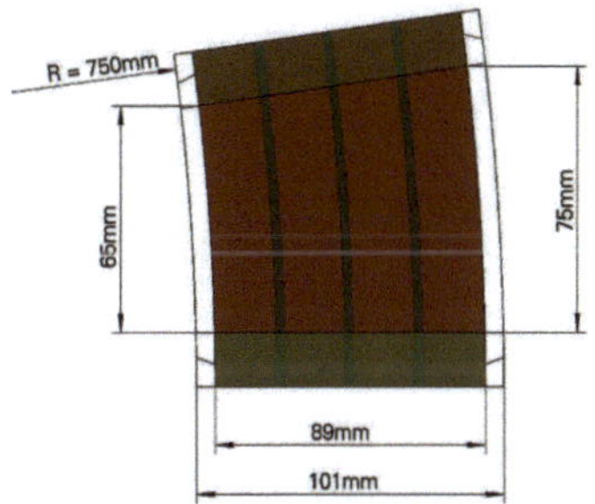

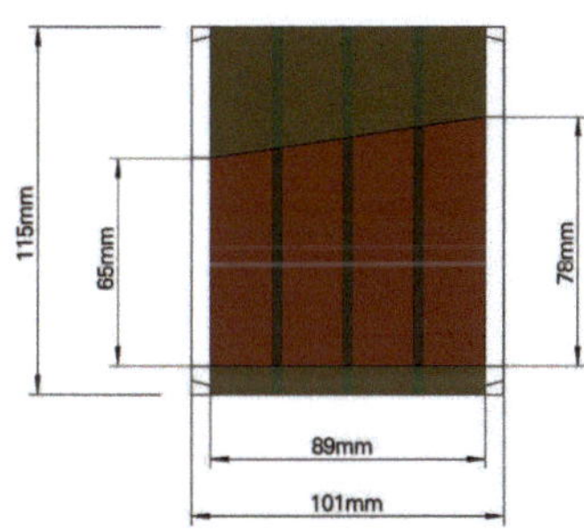

b. Winding geometry before modification   c. Geometry of the winding model

Figure 4.2: Symmetric section in winding and the geometry of the winding model.

The thickness of the wrapping papers, used around conductors, has a crucial impact on the heat transfer between the conductors and their surroundings due to the different thermal characteristics. However, precise and adequate insulation layers are also challenging to achieve, especially when dealing with models that have relatively short conductors. Typically, single conductors in HV-windings are insulated with paper because preferable insulation properties to oil impregnated paper.

The applied thickness has substantial impact of the heat transfer mechanism between the conductors and surroundings, due to the respective thermal properties. Therefore, achieving precise and well-defined thickness of the insulation can be challenging. Due to the resulting uncertainties, the wrapping papers of the conductors are not included into the experimental design. Instead, PVDF strips are employed between the conductors to provide similar thermal conductivity in the range of oil impregnated papers within the specific dimensional ranges.

Finally, the flow meter, the flow heater, the gear pump, and a control unit for supplying power to each heating cartridge are inserted to obtain the specific boundary conditions. It allows accurate control over the mass flow rate $\dot{m}_{oil}$, the oil inlet temperature at the model T_{in} and the heating source of each conductor. The pipeline and instrumentation diagram applied in the experimental setup follows the set up in [110]. The primary purpose of examining experimental winding model to collect measurement data to validate numerical CFD results.

Since it focuses on winding thermal aspects, the model allows temperature measurements at every conductor element using a PT-100 temperature sensor. Such temperature measurement can be used to identify hot spot temperature and its location. Besides, model A has been validated with the measured results in [110]. This chapter focuses on evaluating the influence of various geometrical parameters, including the height of the horizontal channel, the width of the vertical channel, and the number of barriers in the windings. Additionally, it discusses operational conditions, including the rate of heat losses at different inlet flow rates. The optimized design provides a lower average temperature for conductors and keeps the average temperature lower at different operational conditions to avoid thermal stresses. Additionally, the experimental and numerical design considers the influences of non-uniform power losses.

Since the losses at each single conductor element are applied separately, the experimental model has shown possibilities for defining those operating parameters. Fluid guides (washers) mainly force the oil to enter a pass through only one vertical channel and leave the pass via the opposite channel, leading to a zig-zag flow distribution. Moreover, the washers cause more consistent flow distribution inside the horizontal channels and, subsequently, enhance the heat transfer on the disc surfaces.

4.2 Numerical Investigation of the Affecting Parameters

Table 4.1 shows the values of geometric parameters that are investigated.

Table 4.1: Values of the investigated affecting parameters.

Affecting Parameters	Values
Height of Horizontal Channels (H)	4 mm, 6 mm
Width of Vertical Channels (W)	6 mm, 10 mm
Number of Discs per Pass	6, 10
Inlet Oil Temperature	80 °C

The height and the width of the horizontal and vertical channels are selected based on the electrical specifications.

In contrast to other parameters of the winding geometry, the number of discs in a pass is not predetermined by the electrical specifications or standards of disc type winding transformers [110]. Therefore, the influence of this parameter on the thermal aspect of the windings is highly interesting. Consequently, the number of discs in each pass is set at 6 to 10 based on the usual existing disc type winding transformer designs [110]. During the investigations, the applied oil flow rates are maintained at 18 kg/s and 3 kg/s by the inlet oil temperature of 80 °C. Table 4.2 provides the dimensions of the specified models with 6 discs per pass. To clarify the differences between the selected models during the numerical investigation.

Table 4.2: Specification of the models with 6 discs per pass.

Model	Width of vertical channel (mm)	Height of horizontal channel (mm)
6W-4H	6	4
6W-6H	6	6
10W-4H	10	4
10W-6H	10	6

An increase in the number of discs from 6 to 10 in a passage also leads to an increase between the lowest and the highest share of oil. Figure 4.3 shows the share of the oil within the horizontal cooling channel at the flow rate of 18 kg/s for 10 discs per pass. Figure 4.4 indicates the share of the oil at the flow rate of 18 kg/s for 6 discs per pass.

Additionally, the higher number of subsections within the horizontal oil channels divided by washers leads to more share of oil inside the channels. This is evident in the zig-zag design using 6 discs. Moreover, the diagrams show the influence of the height of the horizontal channels. An increase in the height of the horizontal channel can be evaluated while keeping other parameters unchanged. This causes a decrease in the share of oil in the lower horizontal channels and results in non-uniform oil distribution.

This phenomenon increases the number of flow eddies at the entrance of channels which can be explained through visualization. It is also visible that the middle section of the pass with 10 discs includes the lower share of the cooling oil. Figures 4.5a and 4.5b demonstrate the temperature gradients using different geometrical designs at two different inlet flow rates, 3 kg/s and 18 kg/s, respectively. The temperature at each conductor is compared with the oil inlet temperature to visualize the gradient of the temperatures.

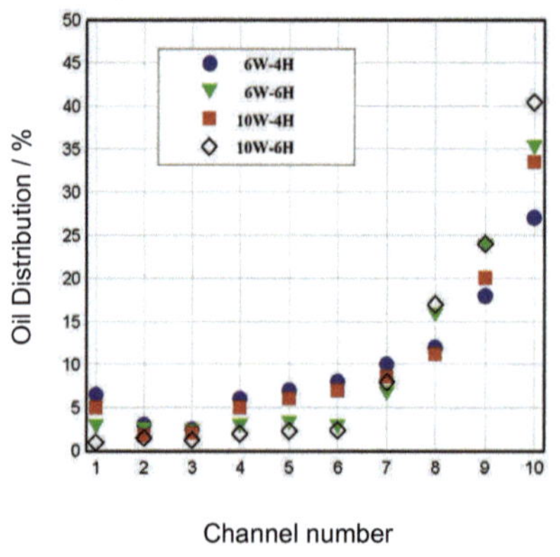

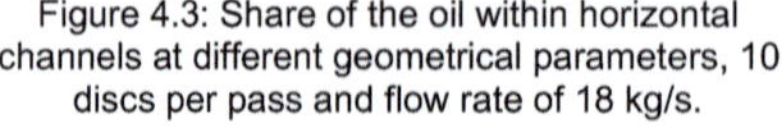

width of vertical channels (W), height of horizontal channels (H)

Figure 4.3: Share of the oil within horizontal channels at different geometrical parameters, 10 discs per pass and flow rate of 18 kg/s.

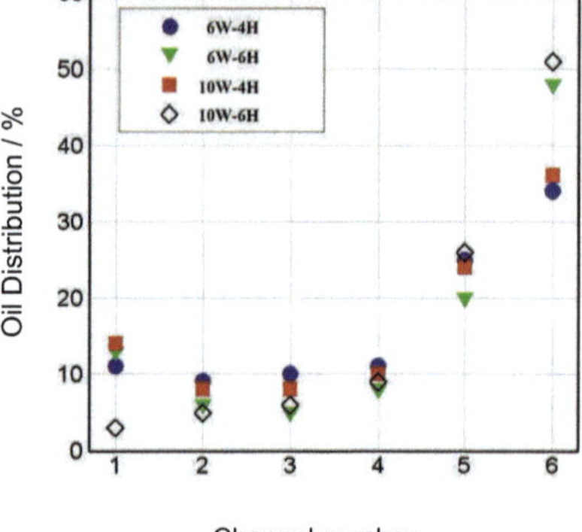

width of vertical channels (W), height of horizontal channels (H)

Figure 4.4: Share of the oil within horizontal channels at different geometrical parameters, 6 discs per pass and flow rate of 18 kg/s.

Each disc has 4 conductors with 6 discs per pass, representing the temperature of 24 different conductors for each model per graph. The design, including the height set to 4 mm and the width set to 6 mm, provides lower temperature gradients at both inlet flow rates and the higher flow rate within the model provides the lower average temperature of the conductors.

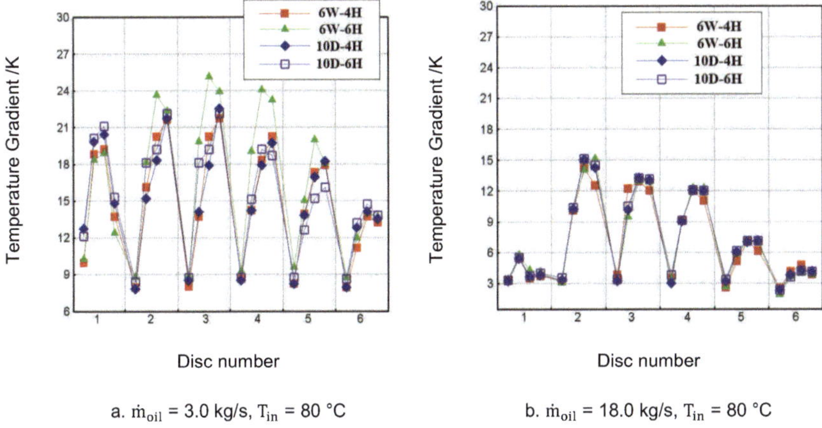

a. $\dot{m}_{oil}$ = 3.0 kg/s, T_{in} = 80 °C   b. $\dot{m}_{oil}$ = 18.0 kg/s, T_{in} = 80 °C

Figure 4.5: Temperature gradient using different geometrical designs at two different inlet flow rates.

Figure 4.6 shows four different geometrical designs with 6 discs per pass to illustrate the temperature contours at the flow rate of 3 kg/s. It is evident at the inlet is located on the right side of each contour, and the fluid flow is blocked on the left side to provide the oil-directed cooling behavior.

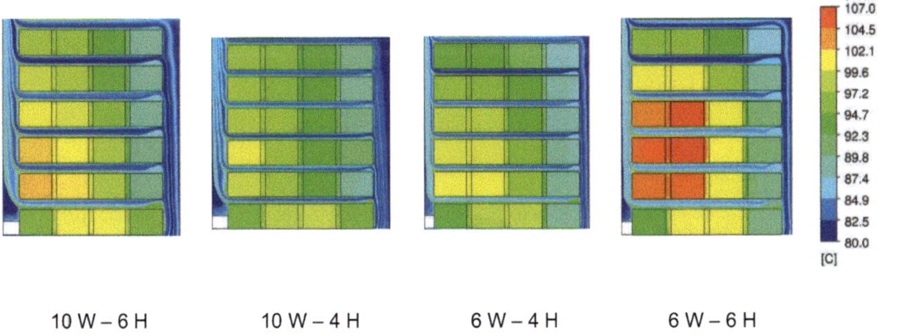

Figure 4.6: Temperature distributions within the winding model at $\dot{m}_{oil}$ = 3 kg/s, T_{in} = 80 °C.

The hot streaks are created along the vertical channels and flow upwards with the higher temperature. Flow stagnation is observed at the inlet of the horizontal channels. Overheated local sections emerge within the winding at the location of stagnation. The heat losses are ineffectively transported by the circulated oil at the stagnation area. After analysing the thermal conditions of four different geometrical designs, two models representing the highest and lowest average temperatures in the winding. The streamlines and the temperature distribution within the domains are shown in Figures 4.7 and 4.8.

In Figure 4.7a, the streamlines and the velocity distribution are illustrated for model 10W-6H simultaneously; meanwhile, in Figure 4.7b, the temperature distribution for model 10W-6H and the location of the hot spot is given. The same numbers of referenced points are uniformly distributed over the inlet models for visualization at the streamlines.

The flow eddies lead to non-uniform oil distribution at the inlets of horizontal channels. The magnified views of the outlets are shown for a better interpretation of the results. As observed, the oil flow through the second channel redirects towards the first channel instead of rising upwards. The heat transfer process is impaired since the oil enters the first horizontal channel through both the inlet and the outlet. The hydrodynamic arguments can explain the mechanism of creating local overheating and the reversed flow.

Widening the vertical channel facilitates the easier rise of the oil level. This, in turn, leads to an increased oil flow in the upper three channels, accompanied by a more pronounced non-uniform oil distribution. The reversed flow causes a significant local overheating at the second disc in the pass, which becomes the hottest disc within the winding.

The study also gives a hot spot at the second conductor of the disc above the washer (disc 2). Moreover, the outlet region of the lower horizontal channels shows a stagnant flow (area without visible streamlines) or a reversed flow in the first horizontal channel of a pass. Figure 4.8a simultaneously illustrates the streamlines and the velocity distribution for model 6D-4H. In Figure 4.8b, the temperature distribution for model 6D-4H and the location of the hot spot is given. The height and width of the channels are reduced to 4 mm and 6 mm, respectively.

The outer vertical channel acts as an inlet for the horizontal channels at the considered pass, whereas the inner vertical channel acts as an outlet. This configuration leads to lower flow eddies and, thus, a reduction of non-uniform flow distribution.

Further investigations carry out that the height of the horizontal channel is the primary factor in flow eddies. Moreover, in design 6D-4H recorded a hot spot temperature that is 6 K lower. The hot spot is located at the first conductor of disc 2 in this design. The study observes that reducing the height of the horizontal channels also decreases the total size of the winding, allowing better cooling with less volume of oil.

However, if the selected values for the height of the horizontal channels are too small, the risk of unforeseen manufacturing tolerances cannot be neglected. Hence, from a manufacturing perspective, it is advisable to avoid using narrow gaps for the oil path. To determine the influence of using 10 discs per pass, Figure 4.9 illustrates the temperature distribution and the streamlines of the winding model at the inlet temperature of 80 °C and flow rate of 3 kg/s.

Due to the washers on the left side, the oil flows from the right side to the pass and is distributed within the horizontal channels. Afterwards, the oil leaves the pass from the left side to the next pass, representing the zig-zag cooling model. The eddies are emerged at the inlet of the horizontal channels; consequently, the share of the oil at the middle horizontal channels is lower than the last upper channels.

This phenomenon leads to a higher temperature in the conductors located in the middle region of the investigated pass. Decreasing the size of the horizontal channels from 6 mm to 4 mm reduces the eddies, and the conductors are cooled better. This observation is attributed to the more uniform oil distribution through the channels. Using this design (H= 4 mm), the hot spot temperature is 6 K lower than the latter (H= 6 mm) and the average pass temperature within the winding model is lower.

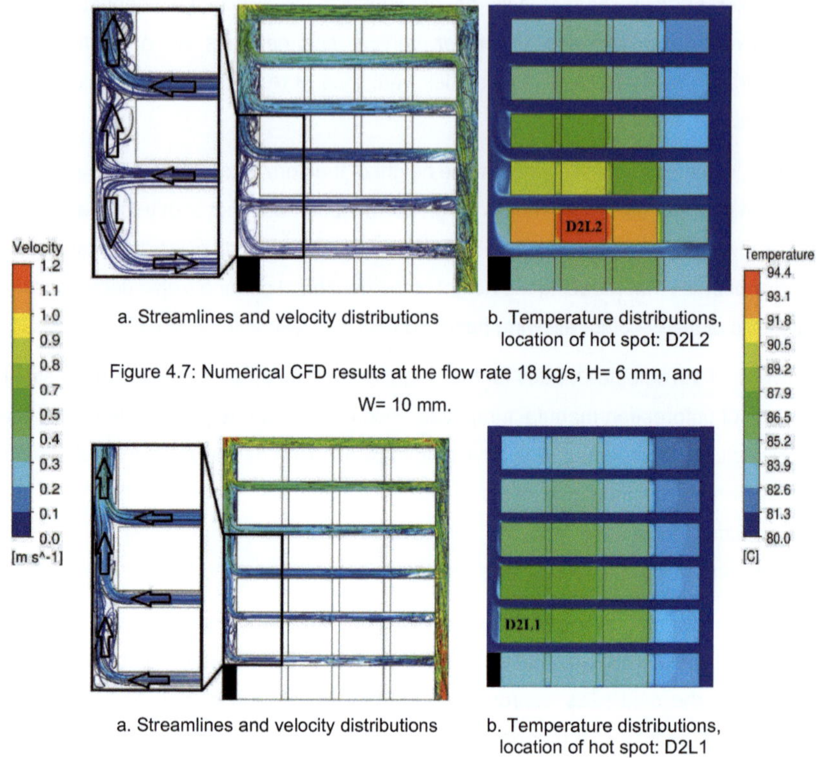

a. Streamlines and velocity distributions   b. Temperature distributions, location of hot spot: D2L2

Figure 4.7: Numerical CFD results at the flow rate 18 kg/s, H= 6 mm, and W= 10 mm.

a. Streamlines and velocity distributions   b. Temperature distributions, location of hot spot: D2L1

Figure 4.8: Numerical CFD results at the flow rate 18 kg/s, H= 4 mm, and W= 6 mm.

However, the position of the hot spot is not changed by the different heights of the channels. As shown in Figure 4.9, the hot spot temperature in both designs (H= 4 mm and H= 6 mm) is located at D6L2). Table 4.3 presents the average temperature distribution (AT), hot spot temperature (HST) and the location of the hot spot (HSL) at two different washer positions. To calculate the average temperature of the conductors in the winding, the temperature of each conductor is determined through CFD calculations.

These individual temperatures are then averaged to obtain the mean temperature, which is essential for discussing heat loss factors. It illustrates the various geometrical conditions with inlet temperatures of 80 °C and a flow rate of 18 kg/s.

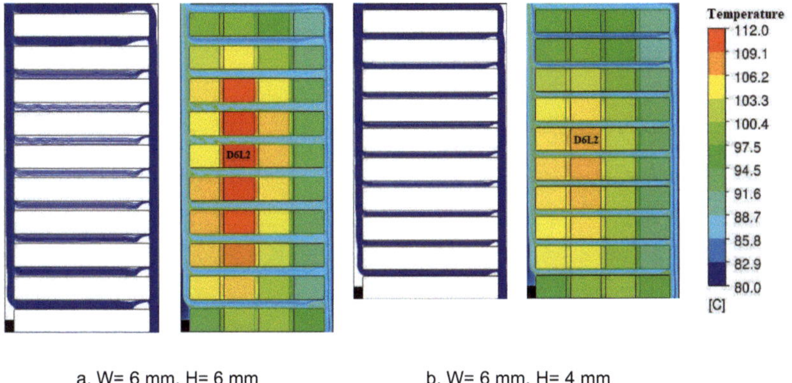

a. W= 6 mm, H= 6 mm   b. W= 6 mm, H= 4 mm

Figure 4.9: Streamlines and temperature distributions in different geometries at oil mass flow of 3 kg/s and 10 discs per pass, inlet oil temperature 80 °C.

In the table, „D" represents the disc number in the winding and „L" represents the conductor number. 10 discs per pass are compared with 6 discs per pass to show the effect of the washers in the zig-zag disc type design. Fewer discs between two successive fluids decrease the hot spot and average oil temperature due to the higher share of the oil, divided within the fewer horizontal channels.

Furthermore, the design with 10 discs and the height of the horizontal channel at 4 mm shows a lower average temperature within the winding. The oil raises upwards to the last two horizontal channels of the pass at the larger vertical channel (10 mm) by 10 discs per pass design. This allows more share of oil in the upper channels so that the hot spot is located at the bottom discs.

As there is a higher oil pressure drop for the design with 6 discs per pass, this study recommends designing 6 discs per pass for higher oil flow rates in oil-directed cooling. This study infers that the optimized cooling condition can be achieved with the height of horizontal channels of 4 mm and the width of vertical channels of 6 mm using 6 discs in pass.

It should be noted the lowest average temperature (AT) and sufficient heat transfer represents the optimized winding designs. The effect of geometrical parameters shows that the lower number of discs within a pass in a zig-zag cooling type leads to a uniformly distribution of the cooling liquids inside the horizontal channels.

Table 4.3: Overview of the numerical CFD results with the different geometrical parameters at the mass flow rate of 18 kg/s and inlet temperature of 80 °C.

Geometrical Parameters of Winding Model (mm)			Hot-spot Temperature (HST) (°C)	Averaged Temperature (AT) (°C)	Hot-spot Temperature Location (HSL) (Conductor- Nr.)
6 Discs per pass	D= 6	H= 4	88	83.2	D2L2
		H= 6	91.1	84.9	D2L2
	D= 10	H= 4	90.7	84.0	D2L2
		H= 6	94.4	85.5	D2L2
10 Discs per pass	D= 6	H= 4	91.7	88.6	D6L2
		H= 6	96.4	91.5	D6L2
	D= 10	H= 4	94.5	89.2	D4L2
		H= 6	97.5	92.9	D4L2

4.3 Influence of Power Loss Distribution

After discussing the influence of geometrical designs on the thermal condition of the winding, this section is focused on the operational condition. The main reasons of heat loss distribution are introduced in Chapter 2, and the reasons for having non-uniform heat dissipation are considered. Investigating the significant effects of the different power loss distributions within the winding is necessary. Due to the non-uniform heat loss distribution within the winding, this study is also visualized the temperature distributions at various heat loss distributions. Furthermore, the influence of different heat loss distributions within the winding model equipped with an additional vertical cooling channel will be determined in the next section.

The top section of the winding is more critical due to the higher temperature of the cooling liquids and the non-uniform heat loss distribution due to the radial leakage flux.

Knowing the eddy current at all positions within the windings is necessary to accurately calculate non-uniform heat losses. It can be set as the initial heat source for each conductor [7]. The comparison between measured winding temperatures with corresponding numerical results at different heat loss distributions are presented to prove the accuracy of the 3D numerical CFD results. Next, the heat sources are described and a detailed view of the numerical models is provided with appropriate boundary conditions.

Heating cartridges have been installed into each conductor at the experimental winding model to apply the electrical heat losses within the winding turns. The heating cartridges are powered by a system consisting of a single DC voltage source, making it feasible to apply different rates of heat losses in the conductors. To ensure the amount of oil flow rate at the model entrance, a digitally controlled valve and a flow meter are employed to provide a specific oil flow rate inside the winding model. The inlet temperature is set via a thermostat controlling the flow heater. While the thermostat provides the sensor signal directly to the heater. Additionally, the study uses two adjacent sensors to monitor the temperature at the entrance. Proper control of the main operating parameters is applied to the laboratory setup to investigate the effects of the non-uniform loss distributions within the winding model.

In this study, the inlet temperature is set to T_{in} = 80 °C while the inlet oil flow rates are set to $\dot{m}_{oil}$ = 3 kg/s and 18 kg/s, representing the laminar and turbulent flows for experimental investigations and the numerical CFD modelling. Furthermore, the distribution approach of heat losses directly affects the thermal aspects of the winding. Since the generated electrical heat losses inside windings are distributed non-uniformly, detailed investigations on the HTS, the HSL, and AT in a winding model at different operation conditions can provide a remarkable outlook for designers.

Thermal boundary conditions are assigned according to the experimental setup conditions, allowing the fluid properties to vary by temperature. 32 W heat loss is integrated uniformly inside each disc (8 W per conductor) at the winding model, and three different heat loss ratios are provided to evaluate the influence of uneven heat loss distributions. The loss at each disc is determined by calculating heat loss factors within HV-windings, as shown in [1]. Eddy current losses cause the differences of the heat losses within the winding. Therefore, at the selected pass, the heat losses are directly depending on the densities of the eddy current losses.

Table 4.4 shows four different conditions for the distribution of heat losses within the winding model of the power transformer, tagged as dimensionless heat losses factor Q_{eddy}. The uniform heat loss distribution (Q=1) and three different non-uniform heat loss distributions (Q_{eddy} = 1.5 – 2 – 2.5) are investigated.

Table 4.4: Details of heat loss distribution within the discs given at selected pass in (W).

Q_{eddy}	Disc 1	Disc 2	Disc 3	Disc 4	Disc 5	Disc 6	$Q_{ave/pass}$
1.0	32	32	32	32	32	32	32
1.5	32	35.2	38.4	41.6	44.3	48	40
2.0	32	49.2	51.6	53.7	59.3	64	51
2.5	32	58.7	64	69.3	73.8	80	63

As depicted, there is an increase in the heat losses at the top discs due to the increased contribution of the radial leakage flux at the end of the winding. To show the ratio of the heat losses in different turns, L1 corresponds to the first layer near the core of the winding, with the number of layers increasing up to L4 in the horizontal direction. The rate of losses is based on data from actual power transformer conditions [1].

Table 4.5 presents the distribution of losses over the conductors of discs toward the radial direction at the last pass at Q_{eddy} = 2, while Table 4.4 describes the different densities at each disc from top to bottom of a pass. The average rate at the first disc of the heat losses is observed at Q_{eddy} = 2. Notably, there is an increase in eddy losses for the top discs due to the greater contribution of radial leakage flux at the ends of the windings, as calculated in [1]. It should be noted that leakage fluxes are higher at the top of the windings, which increases the losses for the uppermost pass at the top of the winding.

Table 4.6 presents the results of the experimental measurements and CFD for the non-uniform heat loss distribution. The flow rate of 3 kg/s is applied for this investigation. The location of the overheated conductor is observed at the third turn of disc 3 and disc 4. The outer and inner turns have an extra cooling surface at the vertical direction, making it possible to cool the conductor at higher heat loss factors. The average temperature of the winding is calculated by the mean value of the temperatures at the pass for both numerical and experimental approaches.

Table 4.5: Heat losses per turn at $Q_{eddy} = 2$ at last top pass. (At the uniform condition 8 W /turn.)

Heat Losses (W)	L4	L3	L2	L1	Sum
$Q_{disc\ 6}$	14.48	14.96	16.4	18.16	64
$Q_{disc\ 5}$	13.52	14	15.28	16.56	59.3
$Q_{disc\ 4}$	12.32	12.56	13.44	15.44	53.7
$Q_{disc\ 3}$	12.16	12.24	13.12	14.16	51.6
$Q_{disc\ 2}$	10.72	11.44	12.96	14.08	49.2
$Q_{disc\ 1}$	8	8	8	8	32

The validated 3D CFD model is used to investigate the temperature distribution in power transformers. The comparison between the CFD results and the measurement data for two different rates of heat losses (Q=1 and Q=2) is illustrated in Figure 4.10.

As shown in Table 4.4, the first disc in both investigations has the same rate of losses (32 W) and an increasing trend to the last disc. Moreover, the temperature gradient at the vertical direction of the diagram in Figure 4.10 refers to the difference between the temperature of each conductor and the inlet oil temperature, at the flow rate of 3 kg/s.

Table 4.6: Details the temperature distribution and hot spot location at the non-uniform and uniform heat losses at the flow rate of 3 kg/s.

Investigation Method	$Q_{eddy} =1$			$Q_{eddy} =1.5$		
	HST (°C)	HSL	AT (°C)	HST(°C)	HSL	AT (°C)
Measured	98.9	Disc 3	93.7	103	Disc 4	96
CFD	99.4	Disc 4	94.8	104	Disc 4	96.7
Operating Condition	$Q_{eddy} =2$			$Q_{eddy} =2.5$		
	HST (°C)	HSL	AT (°C)	HST (°C)	HSL	AT (°C)
Measured	107	Disc 4	97.7	112	Disc 4	99.8
CFD	108	Disc 4	98.4	113	Disc 4	88.8

The temperature gradient of 24 conductors is illustrated to represent each pass with 6 discs and 4 conductors. The higher rate of heat losses increases the average temperature of the winding and the hot spot temperature. In Figure 4.10, the temperature of the conductors in the third and fourth discs is higher in the experimental approach and CFD investigation. This scenario attributes to the share of the oil through the cooling channels nearby these discs.

Meanwhile, Figure 4.11 presents the distribution of flow at two different 3 kg/s and 18 kg/s, respectively. The heat loss distribution does not affect the oil flow distribution; only the flow rate has a significant effect of the oil flow distribution. It is observed that the discs with higher temperatures (overheated region) are close to the cooling channels with the lowest share of the oil. The last two horizontal channels of the pass include the higher oil flow distribution.

Additionally, the average temperature of the conductors is lower at the last disc due to the higher rate of oil. Moreover, the comparison between the two flow rates show that the lower flow rate of 3 kg/s has a more uniform oil distribution than the flow rate of 18 kg/s, but the velocities of oil along the channels are higher for 18 kg/s.

Figures 4.12 and 4.13 show the temperature distribution at different heat loss distributions for 3 kg/s and 18 kg/s. They illustrate that the hot spot temperature and the location of the hot spot can be changed at various eddy loss factors.

The temperature distributions are displayed at the non-uniform loss factors (Q_{eddy} = 1.5 – 2 – 2.5) and the uniform loss distribution (Q_{eddy} = 1). It should also be noted that the inlet oil temperature within the winding model is 80 °C.

Two flow rates of 3 kg/s and 18 kg/s are selected to demonstrate the temperature distributions. The oil fluid flow rate has a direct influence on the cooling conditions of the conductors.

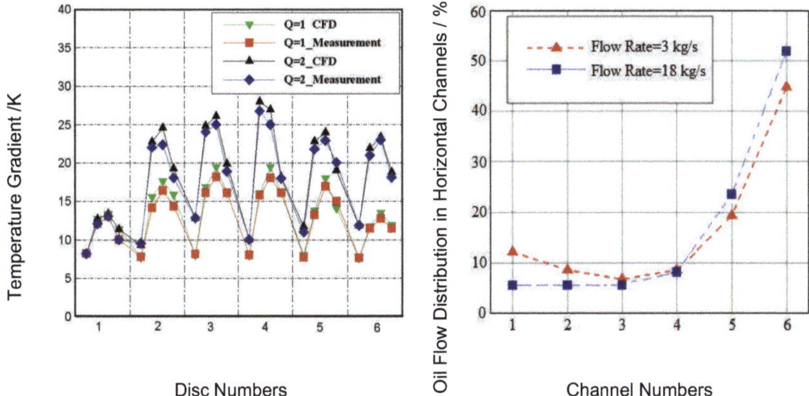

Figure 4.10: Temperature gradient at two different rate of heat losses, inlet oil temperature 80 °C and flow rate 3 kg/s.

Figure 4.11: Percentage of oil flow distribution within the horizontal cooling channels at two different flow rates.

Due to the lower share of the cooling oil at the lower horizontal channels, the average temperature at the middle discs is higher. Considering the last disc, the two middle conductors experience high heat losses. By the non-uniform heat losses, despite the higher heat losses at the top disc of the winding, a hot spot temperature emerges at the middle region of the pass (disc 3 or 4).

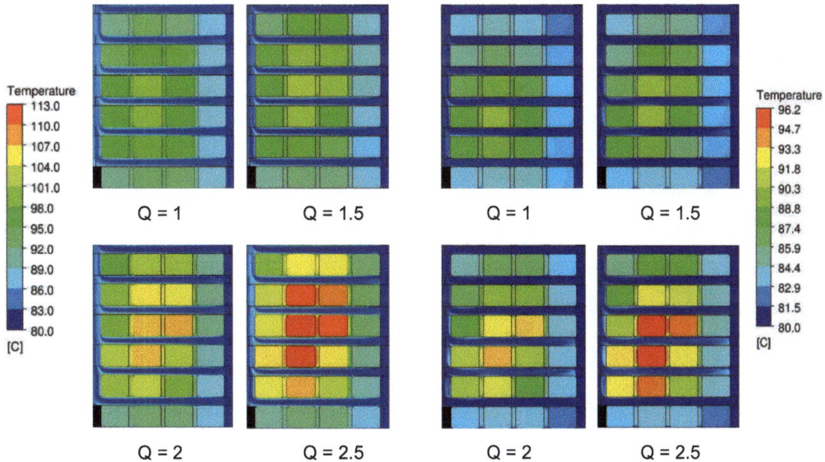

Figure 4.12: Temperature distributions at the inlet flow rate of 3 kg/s and oil temperature of 80 °C.

Figure 4.13: Temperature distributions at the inlet flow rate of 18 kg/s and oil temperature of 80 °C.

The higher share of the oil at the last horizontal channel causes a better heat transfer between oil and conductors. Therefore, despite high heat losses at the top discs, the temperature reduction is noticeable, specifically at the higher heat loss ($Q_{eddy} = 2.5$).

A narrow oil sublayer with high temperature is formed close to the conductors, creating hot oil streaks at the left of the vertical cooling channel. The difference between hot spot temperatures at the flow rates of 3 kg/s and 18 kg/s shows a gradient of 17 K at the same level of heat losses. For the flow rate of 3 kg/s, the hot spot temperature is 113 °C, whereas the hot spot temperature is 96 °C for the flow rate of 18 kg/s. Furthermore, it is observed that non-uniform heat loss distributions directly cause higher average temperatures and hot spot temperatures at both flow rates.

In complement the CFD findings in Figure 4.12 and Figure 4.13, Figure 4.14 offers insight into the relationship between the hot spot factor (H) and the factor (Q) within the numerical results. The concept of the hot spot factor is introduced and the calculation method is explained in Chapter 2. It becomes an evident that H exhibits a linear increase with Q, affirming the presence of a linear relation between the two factors. Besides, the similarity in the slopes of flow rates becomes apparent.

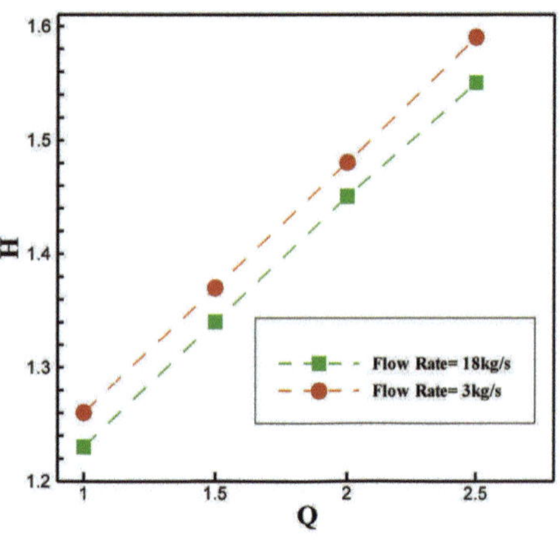

Figure 4.14: Variation of factor H and Q at different flow rates for inlet temperature of 80 °C.

Higher flow rates are indicative of superior cooling efficiency, as demonstrated by the lower hot spot factor at the flow rate of 18 kg/s. As expected, although the highest heat loss ratio is set at the conductor "L1" of the top disc, the hot spot temperature is not located in the last two upper discs of the pass. An additional cooling surface (vertical channel) near the inner conductor transfers the excessively generated heat through the oil flow.

Besides, no localized overheating is expected in the first disc due to the flow stagnation of the oil. It can be concluded that the radial leakage fluxes increase at the ends of the winding, which significantly influences the average temperature and the hot spot temperature in the winding model. Moreover, the location of the hot spot is remained in the middle section which is strongly depends on the oil distributions in the horizontal channel.

4.4 Modification of Geometry Using Additional Vertical Cooling Channel in Model B

This section considers using additional vertical cooling channel to take advantage of geometrical modification. As mentioned in the previous section, the cooling process at the oil channels depends on the oil flow distribution.

As a well-distributed cooling oil within the winding provides efficient convectional heat transfer, it is necessary to determine the influence of different designs on the oil flow distribution. In winding model B, each pass consists of 6 discs, 8 conductors and 6 horizontal cooling channels, with washers to obtain a zig-zag cooling condition. Each conductor is fully covered with an insulating layer, a thickness of 0.6 mm; therefore, the insulation thickness between two conductors nearby is 1.2 mm.

Figure 4.15 shows the selected segment (8° of the entire disc type winding), including highlighted main pass. Figure 4.16 illustrates the detailed front view of the dimensions and the width of vertical channels. The lengths of gaps with additional vertical cooling channel are similar to the vertical channels, and the position of the vertical oil gaps is given. For the design without additional vertical cooling channel, gaps are created using a thick layer of insulation, depicted in blue, which is the same size as the gaps, to keep the geometrical design and winding size similar.

Figure 4.17 shows the position of gaps and the fluid flow directions to identify the geometrical differences. The effect of additional vertical cooling channel on the winding model will be discussed in Section 4.4.1. Meanwhile, Section 4.4.2 discusses the impact of considering non-uniform heat loss distribution in model B.

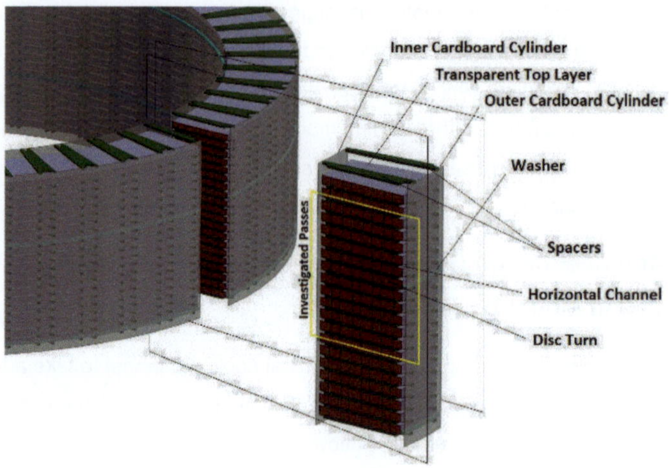

Figure 4.15: Details of the disc type winding model, including the horizontal and the vertical channels, at segment 8°.

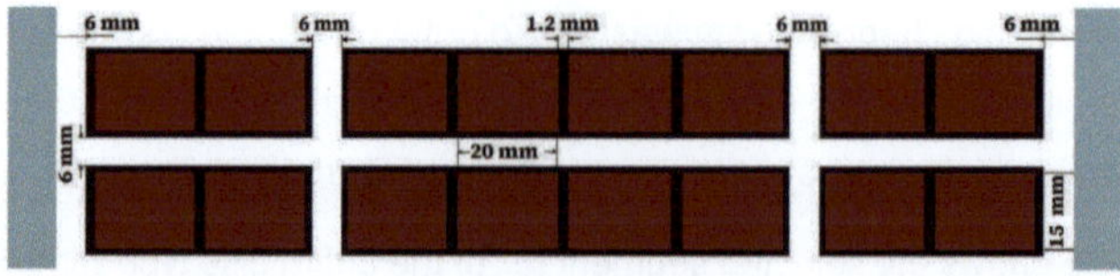

Figure 4.16: Cross-section of the winding model indicating dimensions using additional vertical cooling channel design.

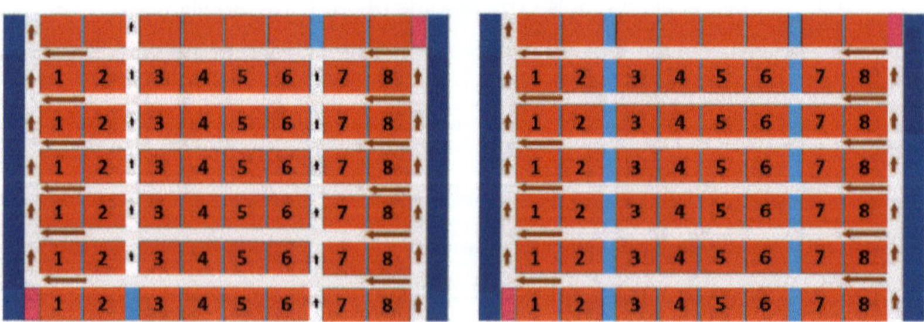

a. With additional vertical cooling channel   b. Without additional vertical cooling channel

Figure 4.17: Detailed view of two different numerical designs at the top pass, including the numbering of conductors and flow directions, 6 discs per pass.

4.4.1 Influence of Additional Vertical Cooling Channel at Uniform Heat Loss Distribution

Figure 4.18 shows the temperature distribution and the velocity distribution within the selected pass of the winding model at the flow rate of 9 kg/s. Uniform heat loss distribution at 8 W/turn is used for both designs, representing 64 W per disc. As seen at same flow rate, the hot spot temperature of the winding model equipped with additional vertical cooling channel is less than 10 K lower than the design without additional vertical cooling channel. Furthermore, the winding model, equipped with additional vertical cooling channel, gives lower average temperature and the temperatures are distributed with a lower gradient between the conductors.

Due to the uniformly oil distribution, the design with additional vertical cooling channel experience no local overheating. Notably, the entrance section of the bottom horizontal channels experience fewer eddies causing smoother oil flow through the horizontal cooling channels. In contrast to the design equipped with additional vertical cooling channel, the design without additional vertical oil path has more eddies at the inlets of the horizontal channels.

Apart from the thermal point of view, the maximum velocity magnitude is lower in the winding model with additional vertical cooling channel. The lower velocity, attributed to the additional inlet of the pass, results in a lower oil pressure difference between the top and the bottom of each pass.

The second conductor of the disc above the washer experience a temperature at about 117 °C, and the hot spot is located at this position for the design without additional vertical cooling channel.

Compared to the other design, the significant effect of a new vertical cooling channel at the right surface of the second conductor is visible. This channel allows the oil to flow near the second conductor, reducing the average temperature in the middle region. The lowest temperature is recorded in the last two conductors of each disc. The heat losses transfer via the convection heat transfer and the conduction of the heat between insulation and conductors are recognizable. The heat dissipation within the insulations between the second and third conductor is remarkably visible at each disc in the design without additional vertical channel.

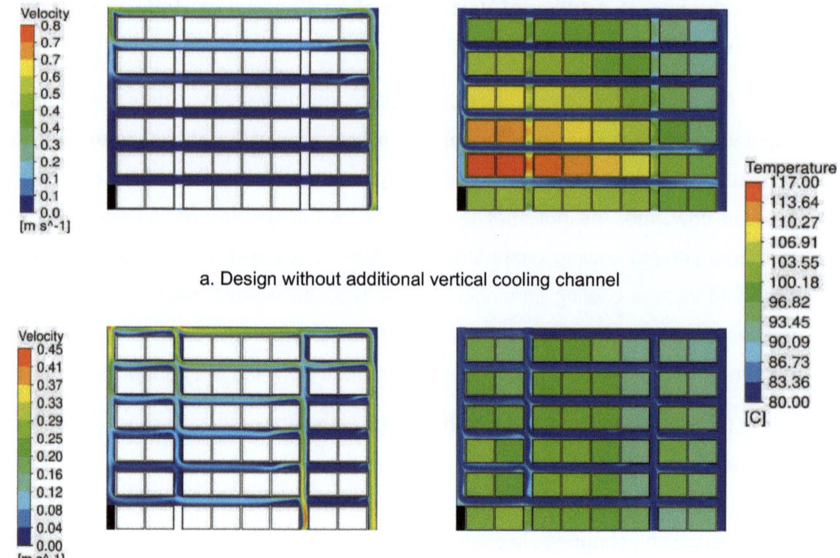

a. Design without additional vertical cooling channel

b. Design with additional vertical cooling channel

Figure 4.18: Detailed view of temperature distributions (right side) and the velocity distributions (left side) at the flow rate of 9 kg/s.

The middle section of the insulation has a higher temperature, and the interface close to the oil has a lower temperature. Considering the oil domain, in the design without additional vertical cooling channel, the boundary layer of oil emerges, which records a higher temperature. The velocity of the oil by design without additional vertical cooling channel at the first two horizontal oil channels is very low. Therefore, the hot spot temperature is located at the second disc. Compared with the design equipped with additional vertical cooling channel, the velocity of the oil flow is higher at all horizontal channels.

4.4.2 Influence of Using Additional Vertical Cooling Channel at Non-Uniform Heat Losses

This section discusses the effects of the non-uniform heat loss distribution on the hot spot temperature of the winding model, equipped with additional vertical cooling channel. In this section, the inlet flow rate is 3 kg/s, and the inlet oil temperature is 80 °C. Uniform heat loss distribution (8 W/turn) is comparable with the heat loss distribution at $Q_{eddy} = 2$.

Figure 4.19 shows the rate of heat losses at the winding per conductor, referred to in the calculation in [1]. The upper disc (disc 6) has higher heat loss, representing the heat loss distribution $Q_{eddy} = 2$.

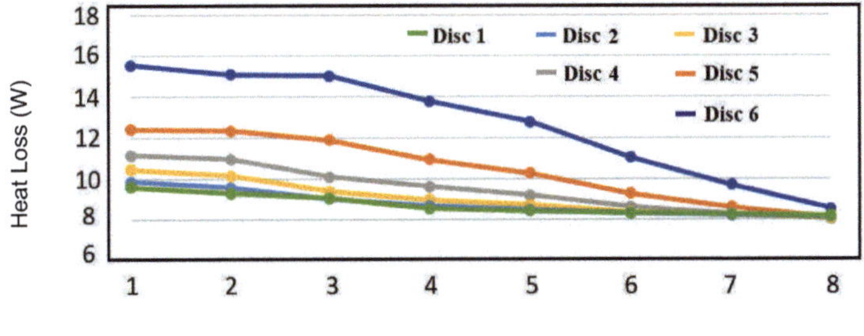

Numbering of conductors per disc

Figure 4.19: Distribution of the heat losses within the pass at $Q_{eddy}= 2$.

In addition, at the same disc, the outer turns of the winding have lower flux leakage rates in the radial direction than the inner turns.

The rate of the heat losses at each pass is obtained from the measured heat loss distribution within the winding of the power transformer. The vertical axis of the diagram shows the heat loss in (W), and the horizontal axis shows the turn no. (conductor no.). The first conductor at the inner side of the winding experiences higher heat losses. Figure 4.20a illustrates the temperature distribution within the pass, equipped with additional vertical cooling channel at uniform heat loss; whereas, in Figure 4.20b, a non-uniform heat loss distribution is set to investigate the temperature distribution with the same geometrical design.

As shown in Figure 4.20a, the hot spot temperature for the uniform heat loss is at 112 °C while in Figure 4.20b, the hot spot temperature for the non-uniform heat distribution is at 134 °C. While both designs have the same structure (equipped with an additional vertical cooling channel), the hot spot is positioned differently. In Figure 4.20a, the hot spot is located at the first disc and the second conductor. The lower oil flow rate through the first horizontal channel leads to overheating region at the first disc. The remarkable influence of the oil flow is apparent at the bottom left side of the pass. The conductors with higher temperatures are in this region due to the lower flow rate.

The thick insulation layer between the second and third conductors at the first disc dissipates the heat from vertical interfaces of the adjacent conductors. Figure 4.20b depicts the position of the hot spot at the last disc (disc 6, conductor 4). Excessive heat losses lead to the higher temperature in this section. The average temperature of the winding is also higher due to the additional rate of losses in the winding. The middle section of the last disc experiences the higher temperature, leads to additional overheating within this region.

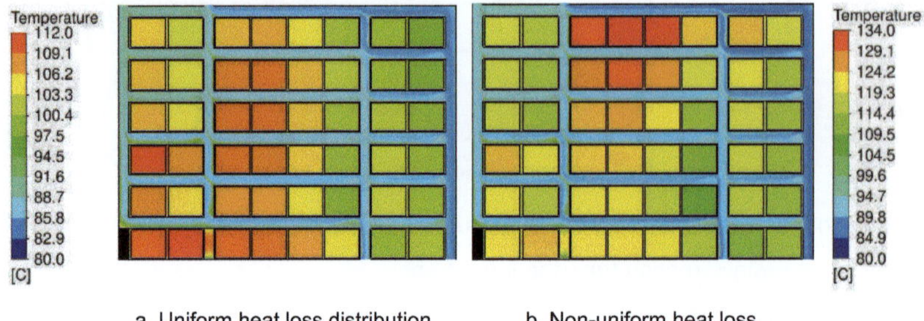

a. Uniform heat loss distribution   b. Non-uniform heat loss distribution

Figure 4.20: Temperature distribution within the pass equipped with an additional vertical cooling channel at the flow rate of 3 kg/s.

The conductors at the right segment of the pass have two additional cooling interfaces, which keeps the local temperature lower at this area. In the design without additional vertical cooling channel, the oil flows upward first through only the main right vertical channel. Then, the eddies at the inlet of the horizontal channels impair the oil distribution within the first horizontal channels.

Despite the higher losses at the last discs, the higher share of the oil keeps the temperature of the discs lower than the bottom section. The generated heat energy is conducted from vertical interfaces and transferred to the horizontal interfaces between the insulation and the oil. The temperature of the oil is increased from the right segment to the left segment, represents that the oil creates hot strikes upwards, directly to the outlet of pass.

Consequently, the optimum combination from a thermal perspective is when the heat distribution is uniform within a winding design using the additional vertical cooling channel. The additional interface of oil and the insulation can lead to better heat transfer along the vertical oil channels.

The study indicates that the design equipped with additional vertical cooling channel provides better cooling performance to avoid eddies at the inlet of the horizontal channels. Despite of the lower eddies in the design with additional vertical cooling channel, the hot spot locates at the top disc because the non-uniform heat losses dominates within the cooling structural design.

The last disc experiences an excessively higher rate of heat losses. Notably, the higher rate of oil flow cannot eliminate the effect of additional heat losses, which is resulted to the localized hot spot temperature.

Following the visualizations in Figure 4.20 for the inlet flow rate of 3 kg/s, Table 4.7 summarizes the hot spot and average temperatures in different scenarios for the inlet flow rate of 9 kg/s. At the same heat loss distribution, the design equipped with additional vertical cooling channel reports a 10 K lower average temperature than the design without additional vertical cooling channel. The difference between the two designs (with and without additional vertical cooling channel) is more pronounced at the non-uniform heat loss distribution.

The average temperature difference is 14 K between the two designs at the non-uniform heat loss distribution. The position of the hot spot temperatures at the flow rate of 9 kg/s is given in column "HSL". As explained, "D" represents the disc and "L" is the turn numbers. The higher oil flow rates lead to more oil flow within the bottom horizontal cooling channels in the pass.

In this disc, the highest rate of heat loss distribution is set so that the heat loss level can dominate other affecting parameters. Meanwhile, in the winding design without additional vertical cooling channel, the average temperature at uniform heat loss distribution is 8 K less than the non-uniform heat loss distribution.

Table 4.7: Overview of the numerical thermal condition of the winding model at the different operating conditions and a mass flow rate of 9 kg/s.

Operating Condition	Winding Design	HST (°C)	AT (°C)	HSL
Non-uniform Heat Loss Distribution (Q_{eddy}= 2)	With additional vertical	109	102	D6L5
	Without additional vertical	122	116	D6L4
Uniform Heat Loss Distribution (Q_{eddy}= 1)	With additional vertical	103	98	D2L4
	Without additional vertical	116	108	D2L2

4.5 Influence of Using Natural Ester Oil

In this section, the OD cooling mode is used to compare the thermal and fluid flow behaviour for natural ester and mineral oil. The numerical CFD results for the hot spot temperature and the oil flow distribution are determined. Figures 4.21a and 4.21b present the temperature gradient of the oil liquids. The diagrams indicate measurement results of the mineral oil which is used to validate the CFD approach. Then, numerical CFD calculations are used to compare the thermal behavior of the ester oil and mineral oil (manufactured by Nynas Nytro Lyra X).

The study shows that in the OD design cooling mode, the natural ester oil has a lower hot spot temperature, and the average temperature of the conductors is lower than the employed mineral oil. At the inlet temperature of 80 °C, the negative effect of higher viscosity in natural ester oil is minimized as higher thermal conductivity provided a better cooling condition in the pass. The positions of the hot spot temperatures are not changed using different oil materials. The upper x-axis represents the numbering of the conductors.

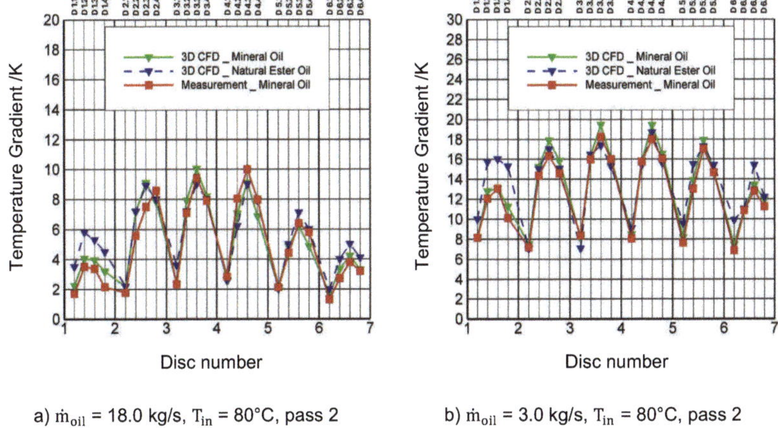

a) $\dot{m}_{oil}$ = 18.0 kg/s, T_{in} = 80°C, pass 2   b) $\dot{m}_{oil}$ = 3.0 kg/s, T_{in} = 80°C, pass 2

Figure 4.21: Temperature gradients between inlet oil temperature of the pass and the conductor for two different oil liquids.

The oil flow distribution for oil liquids is considered. Figure 4.22 shows the share of oil flow distribution (in percentage) within the horizontal channels at two different oil flow rates. In this regard, the oil distribution within the horizontal channels is more homogenous using natural ester oil. Although most of the oil flow rises toward the last two upper horizontal channels, the distribution difference is lower using natural ester oil. For reference, the position of each cooling channel in the pass is given in Figure 4.23.

The temperature distributions of the conductors at the inlet temperature of 80 °C are presented in Figures 4.24a and 4.24b. The hot spot temperature at the flow rate of 3 kg/s is 101 °C in the pass using mineral oil. The eddies at the inlet of the horizontal channels are also visible.

Attending more eddies at the inlet of the horizontal channels, contributing to higher non-uniform oil flow distribution. The hot spot temperature at the flow rate of 18 kg/s is 91 °C. While highest temperature of the pass is not similar for different liquids, the location of the hot spot temperature remains the same. This is attributed to the design of the winding and the lowest oil flow in the third horizontal channel.

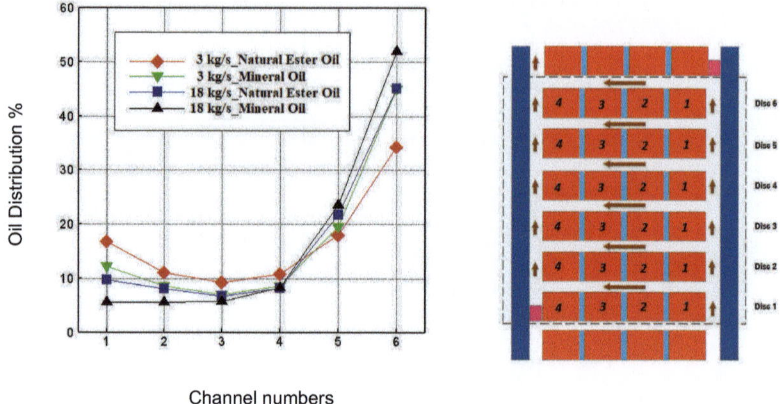

Figure 4.22: Share of the oil flow through channels with different oil liquids and flow rates.

Figure 4.23: Referenced disc numbers in the investigated pass.

It should be noted that although the first horizontal channel has a lower share of the oil, the first disc can be cooled from the bottom interface, connected to the last horizontal channel of the previous pass in the zig-zag disc type winding design. The comparison between two different flow rates, 3 kg/s and 18 kg/s, shows that the higher flow rate in OD cooling mode provides better thermal performance. At the flow rate of 18 kg/s, the average temperature of the winding is lower in both oil liquids. A higher oil flow rate causes a higher convectional heat transfer coefficient which is led to lower average winding temperature.

Figure 4.25 shows the effect of using additional vertical cooling channel with two different oil liquids. The winding design with 8 conductors per disc is employed at the flow rate of 9 kg/s to illustrate the influence on temperature distribution.

The hot spot temperature using natural ester oil is at 99 °C, which indicates better thermal performance in OD cooling mode. The first two discs above the washer records the highest temperature region using various liquid materials. The hot spot temperature in the design with additional vertical cooling channel using mineral oil is 102.5 °C at the flow rate of 9 kg/s.

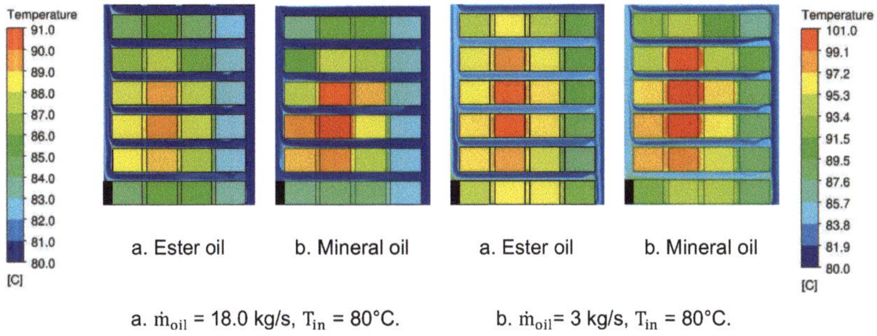

Figure 4.24: Temperature distributions within the investigated pass at different flow rates.

The uniform heat loss of 8 W per conductor (64 W per disc) is contributed to the winding model as the initial heat source for the investigation.

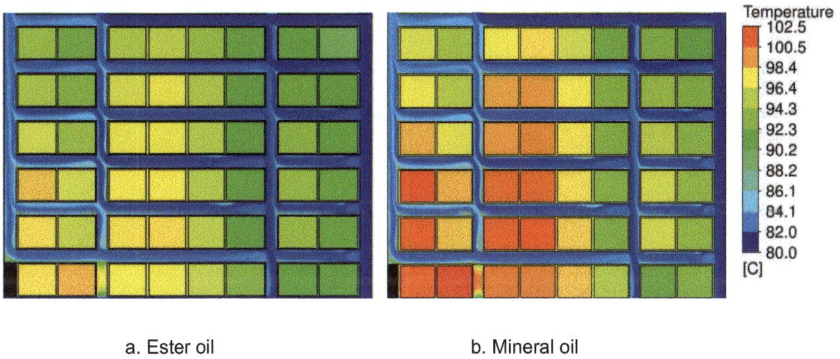

Figure 4.25: Temperature distributions by different oil liquids at a flow rate of 9 kg/s and an oil inlet temperature of 80 °C, equipped with additional vertical cooling channel.

Hence, maintaining a consistent oil flow rate for the winding model and incorporating natural ester oil with an additional vertical cooling channel emerge as the optimal configuration for enhancing the thermal performance of this design when considering OD cooling modes.

5. Investigation of Transient Thermal Behavior within ON-Cooled Windings

This chapter investigates the transient thermal behavior of a distribution transformer during a heat run test and the transient thermal behavior of a winding model in a transparent oil tank. Distribution transformers experience various thermal stresses during their lifetime. As copper losses are the primary heat source inside the transformers and overheating deteriorates the insulation life of a transformer, it is essential to determine the rate of copper losses and monitor the transient temperature distributions. During the test, the total amount of copper losses during full load conditions are determined. Hence, experimental thermal analysis can provide information about the behavior of the transformer.

This study uses a three-phase 400 kVA distribution transformer to perform a short circuit test. The measuring instruments like a wattmeter, voltmeter and ammeter are connected to the HV winding of the distribution transformer. The LV winding is short-circuited, and thermal sensors are utilized to study the behavior of the oil flow and the heat transfer in the domain during the transient condition within the transformer.

Moreover, the transient behavior of the temperature profiles of both the conductors and the mineral oil at the different locations are provided in this chapter. The thermal images are used at different time intervals to see the surface temperatures of the tank and fins. Finally, the highest temperatures of the LV and HV winding at different time intervals are monitored during the test period.

5.1 Experimental Setup

5.1.1 Distribution Transformer Design (Model C)

The test transformer (Model C) is a distribution transformer with a rating of 400 kVA. It has a primary voltage of 10 kV and a secondary voltage of 400 V. The vector group of the transformer is Dy5, i.e., the primary winding is delta connected, and the secondary winding star relates to a shift of 150°. The rated primary and secondary currents are 23.1 A and 577 A, respectively. The HV winding is a disc type winding divided into six stacks with spacers. Each stack consists of 20 discs, with each disc having five turns. The LV has 2 layers, and each layer has 18 turn.

The tank contains Nynas Lyrax mineral oil, and cooling is exclusively accomplished through natural convection, the requirement for an external cooling medium is eliminated.

Figure 5.1 displays the nameplate of the Type HOP 400/10 distribution transformer, manufactured by company Volta- Werke in Berlin, Germany.

Figure 5.1: Nameplate of distribution transformer (model C).

5.1.2 Test Methodology

A short-circuit test is performed on a transformer to measure the full load copper loss in the winding. The LV winding of the transformer is first short-circuited to perform the test. As shown in Figure 5.2, the three phases on the LV side are short-circuited to the neutral point using a thick copper strip. The three phases on the HV side are connected to a three-phase autotransformer that can supply up to 400 V. The short circuit voltage of the transformer applies 3.95% of the rated voltage, i.e., 3.95% of 10kV, equivalent to 395 V. When the voltage on the HV side reached 395 V, the rated current flows in both the HV and LV windings resulting in a full load copper loss in the transformer.

The connections at the top regions of the distribution transformer are given in Figure 5.2. Since the copper loss in the winding causes most of the heat losses in a transformer, a short circuit test can be used to estimate the heat losses at full load. Therefore, the hot spot factor and temperature distribution can be accurately examined according at the rated current level. Figure 5.3 shows the circuit diagram of the short circuit test at the transformer under investigation. The ammeter is connected across the HV winding measures the full load current, while the wattmeter only determines the copper loss in the windings. The iron loss of the transformer depends on the flux. It is less in the short circuit test because of the low value of flux.

The circuit also has a LV side, and the positions of the measurement devices are given. PT-100 sensors are installed in several parts of the transformer, and temperature data is recorded using a PLC temperature data logger every 15 seconds. The location of the installed PT-100 sensors inside and outside the tank can be observed in Figures 5.4 and 5.5. All internal sensors are attached to one winding block of the transformer (phase W). The model is positioned inside the tank and filled with mineral oil which is delivered by the manufacture Nynas.

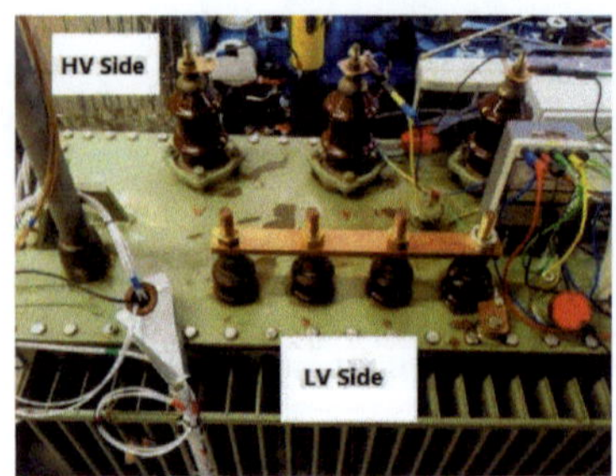

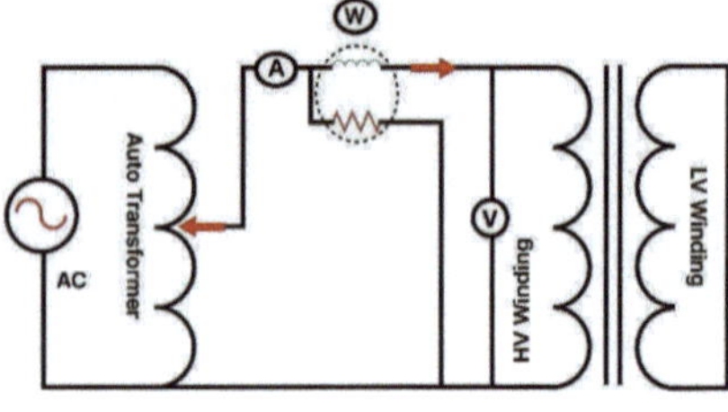

Figure 5.2: Connections on setup, short-circuited LV-side with a thick copper strip.

Figure 5.3: Connection diagram for the heat run test.

The sensors with labels of the windings are as follows:

• Sensor S1 is installed on the outermost conductor of the top stack's 7th disc (from the top). A piece of the paper insulation is removed from the copper conductor so the sensor can measure the copper temperature directly. The sensor is fixed in place using silicone glue, which is used also for remaining sensors.

• Sensor S2 and S3: Both sensors measure the oil temperature at the top and the bottom of the vertical channels between the LV and HV windings. These sensors are placed at the top and bottom without touching any copper or insulation surfaces to measure the oil temperature. The labels of the positions are shown in Figure 5.4.

• Sensor S4 is attached to the LV winding on the rear side of the tank. Since the LV winding is inaccessible from the beginning and the end, the temperature sensor is installed on the conductor that goes to the bushing of phase W.

A part of the insulation is stripped to measure the copper temperature directly, and since the LV conductors in the turns have paper insulation and the conductor going to the bushing only has thin enamel insulation, the part of the conductor with the sensors is covered with additional Nomex paper to form the same insulation as the conductors in the turns, as shown in Figure 5.5.

• The additional sensors are used on the outer side of the tank, attached to the metal surface of the tank. One sensor is fixed about one meter from the tank to capture the ambient temperature during the test procedure. To prevent external oil leakage due to thermal expansion, an attached pipeline is located in place.

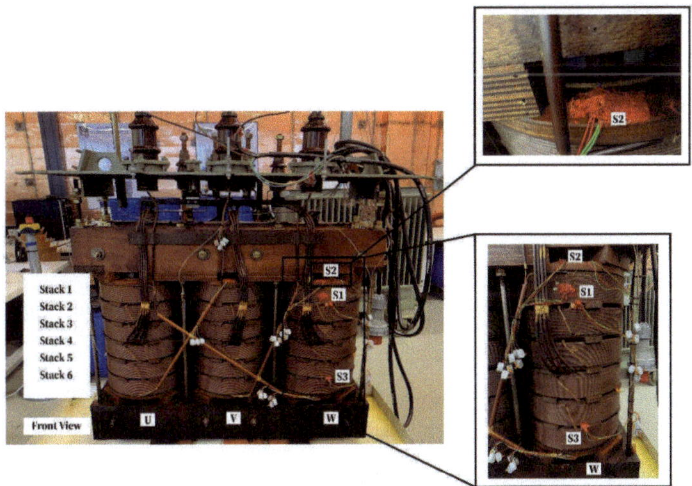

Figure 5.4: Front view of distribution transformer and the arrangement of sensors, phase w.

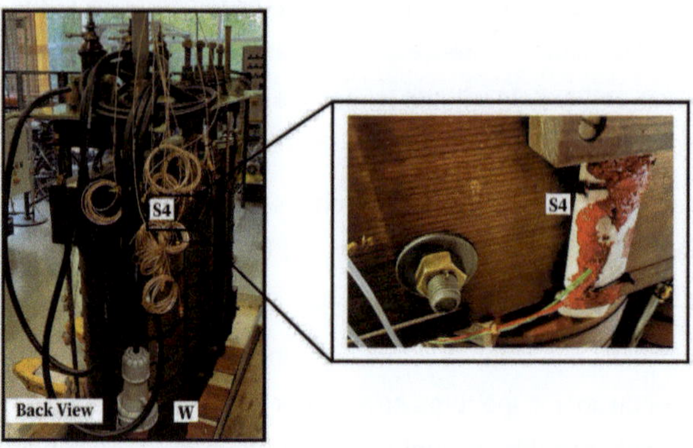

Figure 5.5: Back view of distribution transformer and the arrangement of sensor S4.

It is important to note that the oil mass is maintained constant throughout the measurements, and there is no expansion of oil flow from the tank into the external pipeline. Additionally, the ambient conditions surrounding the experimental setup remain constant without external air ventilation.

5.1.3 Disc type Winding (Model A) within an ON-Designed Cooling System

This section presents the arrangement of the winding model design, which includes four conductors per disc, mentioned in the previous chapter called model A.

The second winding model is constructed using heat cartridges to apply heat losses. Each conductor is equipped with thermal sensors to record the temperature distribution within time intervals during measurements.

Figure 5.6a shows the schematic view of the winding model inside the tank. The zig-zag winding design becomes evident as the fluid flows through the radial cooling channels. The study on natural cooling behavior focuses specifically on the second pass. Figure 5.6b shows the transparent oil tank with the winding model positioned inside just prior to covering the outer walls.

The oil tank has two pipe connections with valves to manually adjust the oil level before measurements. Heat exchanger applies the initial inlet temperature, and the cables are connected to the thermal sensors to measure the temperature of the conductors.

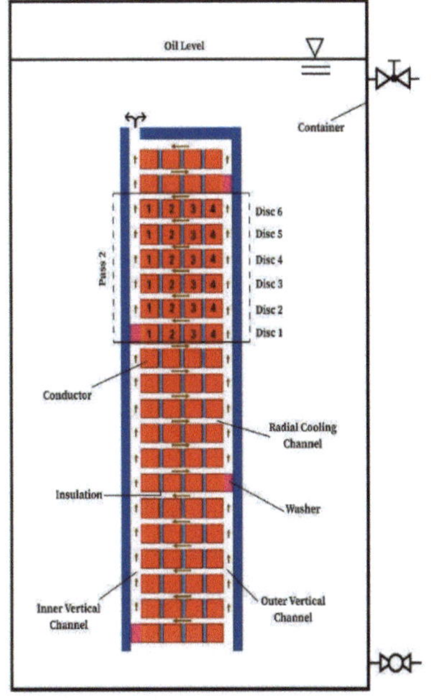

a. Schematic view of the winding model within
the oil tank.

b. Winding model in the transparent oil tank
with sensor connections.

Figure 5.6: Front view of model C before experimental investigations.

The oil temperatures are recorded at the bottom and top of the oil tank. The inlet of
the winding model remains consistently open, allowing the oil to naturally rise due to
thermal forces. Preceding the test, the oil tank is enveloped by a substantial 12 cm-
thick layer of glass wool to minimize heat dissipation into the surroundings. Once the
oil tank is filled with the designated oil liquid at a predetermined temperature, both
the inlet and outlet of the oil tank are sealed. The winding model is also entirely
enclosed. These steps create a closed system, preventing any mass (oil) transfer
and heat transfer to the external environment.

The isolation of the closed system of winding model from the environment and
applying no additional external cooling medium leads to obtain continuous
temperature rise of the oil inside a closed loop.

Therefore, to avoid degrading the insulation materials in the winding model, the maximum temperature of the model during the measurement is imposed to T = 115 °C.

As a uniform heat loss distribution of 8 W per conductor is conducted to the winding model and the starting temperature of the oil is set at 40 °C. The experimental setup allows to perform the measurement at higher temperature of the ambient temperature. Because of no additional cooling media in close to the experimental setup, a low temperature start is not considered in this study.

5.2 Analysis of Transient Thermal Behavior

5.2.1 Distribution Transformer (Model C)

The short circuit test uses mineral oil at an ambient temperature (initial temperature) of 20 °C. Figure 5.7 shows the temperature rise of the numerical CFD modelling and the measurements on the distribution transformer in the laboratory. Since no sensor can be attached to the inner layers of the HV winding, the sensor on the outer layer of the HV winding is used for validation.

Since this measured temperature is not the hot spot temperature, as the location is cooled better by the oil, it records the highest recorded temperature during the short circuit test. It also illustrates the temperature distributions in CFD numerical investigation and the measured data. The conductor temperatures have shown rapid increase in the first two hours. Due to the effective oil circulation, the slope of temperature rise gradually decreases in the next 5 hours.

Figure 5.7 shows two different winding temperatures from the CFD simulations. First, the local temperature of the sensor at the outer layer of the distribution transformer, and second, the hot spot temperature of the winding at the inner layer. At the last step of measurement, the upper oil temperature in the HV winding reaches 72 °C, and the lower oil temperature increases to 63 °C simultaneously. Meanwhile, there is a 9 °C difference between the upper and lower oil temperature. The highest measured temperature in the HV winding is 83 °C and the calculated temperature at the same position is 86 °C.

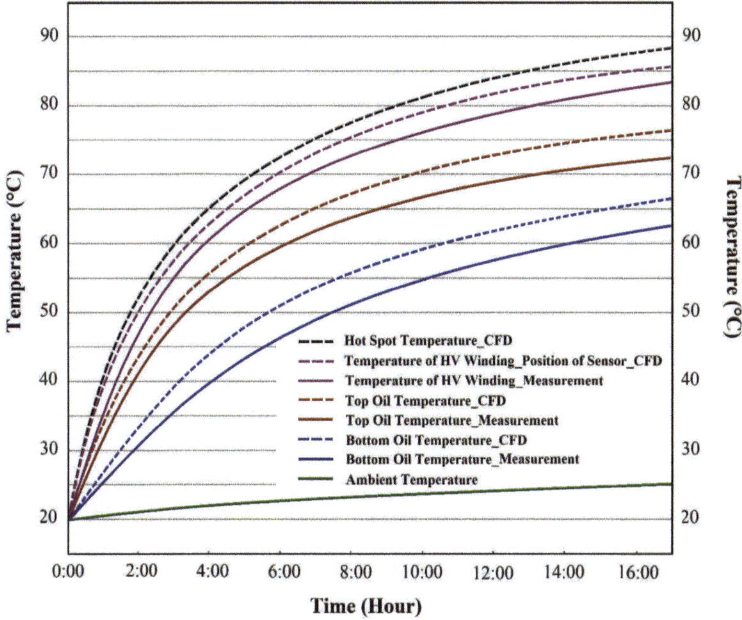

Figure 5.7: Temperature rise in the numerical CFD investigation and the measurements on distribution transformer (model C) at the starting temperature of 20 °C.

Figure 5.8 shows the temperature profiles within the conductors at the starting temperature of 20 °C. After 8 hours, the topmost discs of the upper winding experience higher thermal stresses, as indicated by the temperature distribution. Compared to the rest of the winding, the uppermost stack (stack 1) has higher average winding temperature. After 16 hours, the top area of the HV winding shows a higher temperature with the hot spot temperature at 88 °C.

Although the conductors of the HV winding have the same level of applied heat losses, the upper region indicates the hot spot temperature, and the higher temperature distribution is located at the last stack. Across various time intervals, it becomes evident that temperatures initially rise uniformly across almost all regions. However, as time progresses beyond the time constant, temperature increases begin to vary across different stacks. The time constants of the ON cooling conditions for both models are determined in section 5.3.

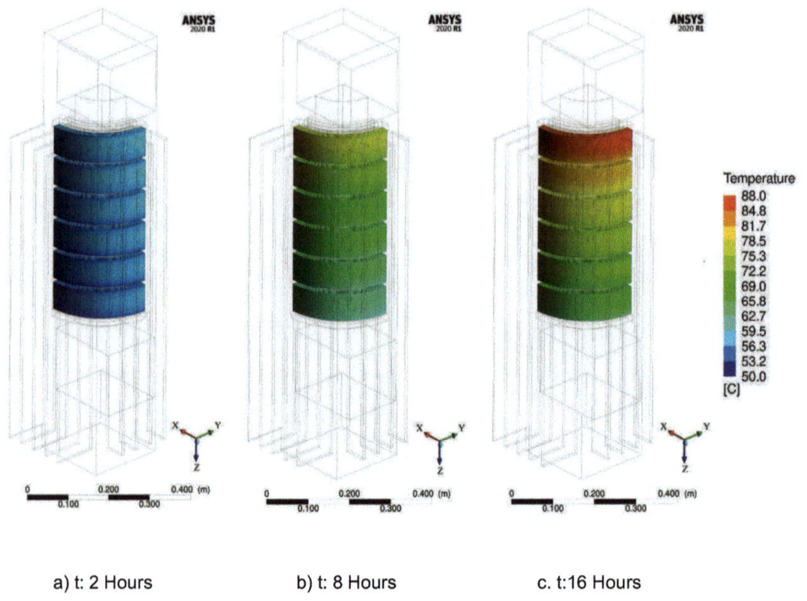

a) t: 2 Hours   b) t: 8 Hours   c. t:16 Hours

Figure 5.8: 3D CFD temperature profiles within the HV winding of the distribution transformer (model C) at different time intervals.

5.2.2 Disc Type Winding (Model A)

The winding model inside the oil container is considered to determine the thermal behavior of an ON-cooled winding model. The uniform losses are given to the conductors, and the time span from the initial moment to t = 6000 s is contributed.

To achieve this, the oil inlet temperature and outlet temperature as a function of time are considered, in combination with the hot spot temperature, as part of the validation process. The temperature of the oil and the conductor are captured at 15 s intervals. In Figure 5.9, the transient temperature distributions in a numerical 3D CFD analysis alongside measurements, starting from an initial temperature of 40 °C. Notably, since there is no cooling system in the oil tank, the rapid temperature rise in both approaches is noticeable.

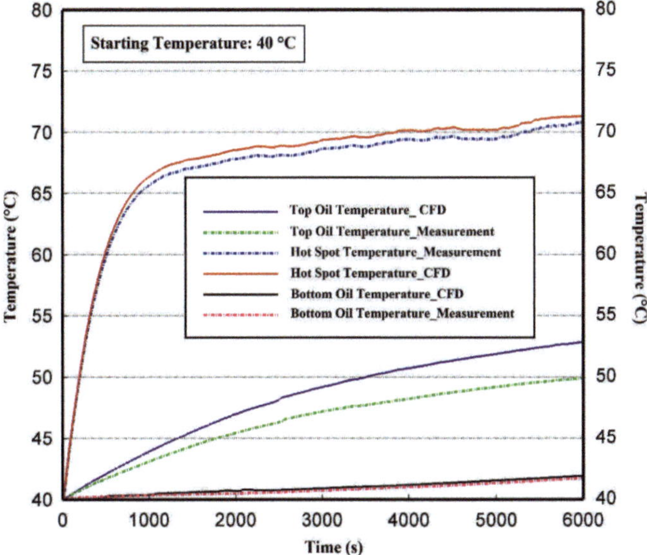

Figure 5.9: Transient temperature distribution within the winding model (model A) using the ON cooling method by starting temperature at 40 °C.

Furthermore, the hot spot temperature in the numerical model closely matches the measurements, deviating by less than 3 K throughout the observation period. During the first 1000 s, the conductors experience very rapidly temperature rise due to the low oil flow velocities in the radial channels. However, after the initial 1000 s, the generated heat is transferred to the surrounding oil, and the oil flows through the cooling channels. This natural convection process within the cooling channels results in a reduced temperature gradient within the conductor.

The rise in the temperature within both oil domain and conductors is notable. Nevertheless, the gradients of the oil temperature at the top and bottom of the winding model are not identical to the rate of increasing of the hot spot temperature. The inlet temperature in the winding model is increased very slowly, with a maximum of 3 K in the entire period measurement due to 150 liters of oil inside the tank. Figure 5.10 illustrates the temperature distribution in the pass in model A, while Table 5.1 presents the precise values of measurement data and CFD results for oil and the hottest conductor.

Utilizing PT-100 sensors in each component of the winding model enables to compute the average temperature in the laboratory. The determination of hot spot factors is achieved by calculating the average temperature of both oil and winding, as elaborated in section 5.3. Figure 5.10 represents the temperature profiles of the winding model alongside streamlines, commencing from an initial temperature of 40 °C. During the initial 100 s, the temperature of the conductors exhibits a gradual increase, mirroring the minor fluctuations in the oil temperature rise. Additionally, the oil velocity does not change considerably during this initial time interval. By the time t= 1000 s has elapsed, the average temperature of the conductors experiences an increase, concurrently reducing the viscosity of the oil due to the generated heat.

This change contributes to a more efficient distribution of oil flow and an increase in the oil velocity within the channels. Since the heat sources are generated inside the winding, viscosity change is initially evident in the extremely thin sublayers adjacent to the conductor surfaces. This effect becomes more pronounced as the oil circulates faster at t= 4000 s, as evident by the more visible streamlines in the radial channels. Since the tank contains approximately 150 liters of oil, the bottom oil temperature changes at lower portion and takes longer time to be noticeable. Furthermore, the oil streamlines within the upper horizontal channels are not visible at first, as reflected by the low flow through these channels. Gradually, the oil begins to circulates and moves upwards through the radial channels.

Moreover, after t = 4000 s, the hot spot temperature is placed at the third conductor of the upper disc. Despite the uniform heat distribution throughout the winding model, the hot spot temperature is within the topmost disc. Local overheating happens due to the lower oil velocity at the upper section of the pass, and since the first and fourth conductors of the disc also have two additional vertical cooling surfaces, the temperature of these conductors remain lower than the middle conductors of each disc. Notably, the highest oil velocity is recorded at the inlet of the pass due to a washer. Furthermore, the first two horizontal channels experience higher velocities, compared to the last two upper channels.

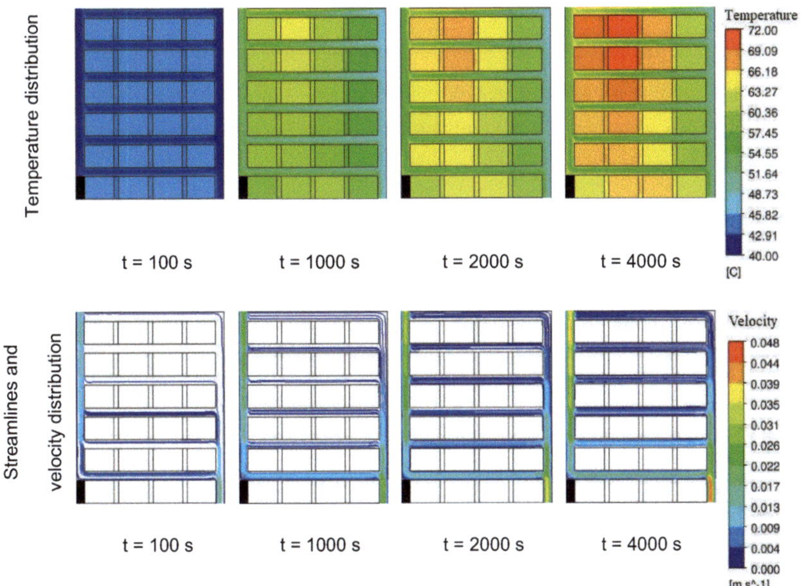

Figure 5.10: Transient temperature and velocity distribution, along with streamlines within the pass, starting at an initial temperature of 40 °C.

It is evident that, under the ON cooling condition, the distribution of oil flow exhibits an opposing pattern, compared to the OD cooling mode. The initial horizontal channels of the pass receive a larger proportion of the oil, resulting in higher oil velocities at the lower region. As illustrated in Figure 5.10, the lower horizontal channels exhibit a higher oil flow rate during ON cooling mode, and the hot spot is placed at the upper disc. Typically, overheating occurs in the final discs during ON cooling conditions. However, when employing the OD cooling mode, the middle discs, or even the initial discs of the pass, experience higher temperatures. Consequently, with the ON cooling condition, the hot spot temperature relocates to the upper disc. The velocity of the oil undergoes changes over time. Table 5.1 indicates the average oil velocity in the channel close to the hot spot temperature.

Table 5.1: Oil velocity within the 5th upper horizontal channel, close to the hot spot location, by CFD calculation at starting temperature of 40 °C.

Time (s)	t = 100	t =1000	t = 2000	t = 4000
Velocity (mm/s)	3.5	9	11	17
Hot spot temperature (°C)	45.43	65.63	67.77	69.42

The trend of increasing average velocity with time is evident. The highest velocity is observed in the first channel, and there is a notable increase in oil velocities in the last two upper channels due to thermal driven forces. It is visible that less oil flows in top cooling channels of a pass causing the hot spot temperature toward the top discs at the uniform heat loss distribution.

5.3 Determination of Hot Spot Factors

Besides the contribution of the transient temperature distributions in both models, this section presents the hot spot factors for both models. The hot spot factor is defined in chapter 2, section 2.1.4. The inclusion of thermal sensors within all the conductors in the disc type winding (model A) facilitates the determination of hot spot factors during experimental investigations. Figure 5.11 depicts the transient hot spot factor, referencing the CFD numerical approach at specific time intervals for the distribution transformer (model C).

One of the main parameters in calculating the hot spot factor is the average temperature of the winding. The CFD investigations involve the computation of both the average oil and winding temperatures. It is worth noting that the initial temperature used for calculating the hot spot temperature is 20 °C in model C. Throughout each time interval, the hot spot temperature is shown on the left y-axis, while the hot spot factor is presented on the right y-axis. The hot spot factor for the experimental approach is not carried out in this model.

The initial two hours exhibit a steeper gradient in the increase of the hot spot temperature. Afterwards, the improved cooling performance resulting from the circulation of oil reduces the rate of temperature rise in subsequent time intervals. Meanwhile, after two hours, the hot spot factor decreases, and a noticeable reduction in the hot spot temperature rise slope becomes apparent in the next time intervals.

Figure 5.12 provides an illustration of the hot spot factors, considering two distinct starting temperatures of 40 °C and 80 °C, within the disc type winding (model A). To ensure the preservation of the insulation system and avoid degradation, a maximum allowable temperature of 115 °C is set during the experimental investigations. Consequently, for both starting temperatures, the analysis of hot spot factors is carried out during the initial 6000 s.

As depicted in Figure 5.12, for both starting temperatures, the hot spot factors at t = 1000 s are higher than those observed in the subsequent time intervals of the investigation. This can be attributed to the fact that during the first t = 1000 s, the oil flow is slower through the cooling channels.

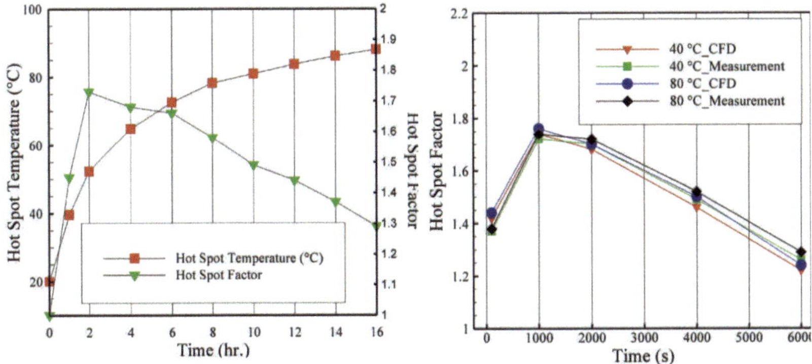

Figure 5.11: Determination of hot spot factor and HST by CFD calculations in model C at starting temperature of 20 °C.

Figure 5.12: Determination of the hot spot factors in model A at starting temperatures of 40 °C and 80 °C.

Following this initial phase, the generated heat within the winding model is efficiently dissipated through convective heat transfer with the circulation of the oil creating an effective cooling cycle without an additional cooling medium. Referring to the establishment of the linear relationship between Q and H in Chapter 2, the noteworthy increase in cooling efficiency (factor S) is attributed to the lower H resulting by uniform heat loss distribution.

It can be inferred that, after a certain time constant within the system, the rapid rise in hot spot temperature diminishes, and the cooling system responds noticeably. Therefore, it is important for designers to figure out the time constant in the transient cooling condition within the winding designs. In this study, the time constants of both models are determined. To achieve this, exponential curves are fitted to the calculated hot spot temperatures by ON cooling condition; then, the time constants are calculated and illustrated in Figures 5.13 and 5.14.

It is evident that model C has a time constant (τ_C) of approximately 4 hours, whereas model A reaches the time constant (τ_A) after 475 s (roughly 8 minutes).

During these time constants, the temperature rise reaches 62.5 % of the maximum temperature increase. Therefore, the calculated hot spot temperature is used for the exponential estimation of the temperature rise. Table 5.2 summarizes the details by illustrating the hot spot and top oil temperatures within model A. It includes the rate of hot spot factors at different elapsed time, which is crucial for assessing the thermal condition of the winding.

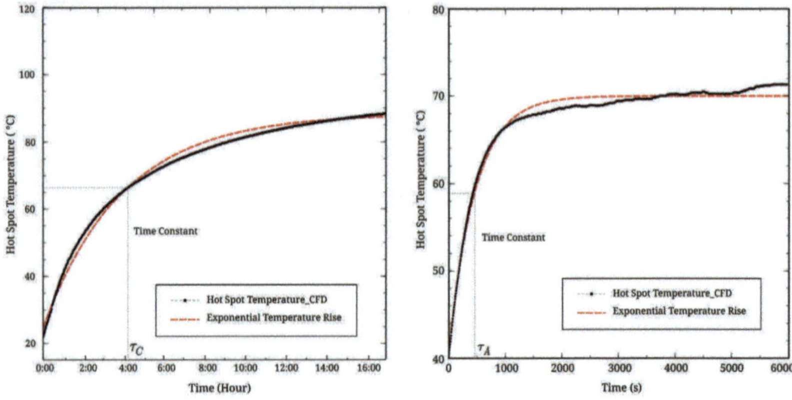

Figure 5.13: Determination of time constant in model C at starting temperature of 20 °C.

Figure 5.14: Determination of time constant in model A at starting temperature of 40 °C.

Both hot spot factors increased rapidly during the first 1000 s. For uniform heat loss, the hot spot factor is directly influenced by both the initial temperatures and the average winding temperatures during the investigation. The results indicate that the hot spot factors with the range between 1.3 and 1.7 in an ON cooled winding model can serve as a basis for assessing the overload capability.

In principle, as shown in [19], the evaluation of factor S is related to the effects of the cooling liquid. For the example with washers, factor S is assumed to be 1. In IEC 60076-2:2011 [19], the H factor is defined through the Q factor and the S factor as H=Q×S. While the H factors are not directly reported in IEC 60076-2, they are calculated using the given Q and S values. Assuming S equals 1, the hot spot factor is directly equal to Q. However, this investigation founds that the calculated S factor is less than 1. For example, under OD condition with H = 1.35 at Q = 1.5, S equals 0.9. This indicates that the S factor is not solely depending on the geometric design but may also vary under different operational condition due to varying hot spot factors.

Table 5.2: Comparison between the numerical CFD results and the measurements at different time intervals in two different starting temperatures, 40 °C and 80 °C.

Approach of Investigation	Parameters	Starting Temperature (°C)	Time (s)				
			100	1000	2000	4000	6000
Measurement	HST (°C)	40	45.67	66.32	68.54	70.15	71.28
		80	85.22	105.1	106.3	108.7	112
	Top Oil Temperature (°C)	40	40.42	43.01	45.45	48.23	49.91
		80	80.26	84.12	90.17	92.72	93.11
	Hot Spot Factor	40	1.37	1.72	1.70	1.49	1.26
		80	1.38	1.74	1.72	1.52	1.29
CFD	HST (°C)	40	45.43	65.63	67.77	69.42	70.02
		80	86.56	107.2	107.9	109.4	113.1
	Top Oil Temperature (°C)	40	40.84	43.93	47.01	50.73	52.83
		80	81.12	86.22	92.06	94.05	96.2
	Hot Spot Factor	40	1.42	1.74	1.68	1.46	1.22
		80	1.44	1.76	1.7	1.5	1.24

In IEC 60076-2:2011 [19], the Q factor ranges from 1 to 1.8, depending on the power rate of the power transformers and the height of the strands. By $S = 1$, the hot spot factor also varies within this range, which is close to the hot spot factor range from 1.3 to 1.7, calculated in this work.

The highlight of this work is that the hotspot factor is provided relative to time, allowing for the evaluation of the thermal condition at any point during operation. The hot spot factors provided in IEC 60076-7 [1] for an ON cooled transformer align with the values, presented in this section, thus enabling a more precise evaluation of overload capability in future applications.

6. Summary and Recommendations for Further Research

Chapter 1 of this work introduces the topic and discusses the motivation behind studying the thermal behaviors of power transformers. Chapter 2 presents the theoretical background of this study, including the governing equations in the fluid flow and heat transfer mechanisms. Additionally, it provides an overview of the state-of-the-art research in the field of thermal analysis in power transformers. It describes that the investigations on thermal performance of the windings can be divided into three main categories.

First, the heat run test, which calculates total heat losses within the windings, while the other two methods use direct measurements and modelling to assess thermal performances. In this regard, these measurements can be carried out using devices like laser-Doppler velocimeters or anemometers to determine fluid flow behavior.

Chapter 2 introduces the concept of thermal modelling, and its main approaches. Moreover, CFD calculations are employed for investigating power the thermal behavior. This chapter also explores advantages and disadvantages of using the CFD method, along with other noteworthy developments in the field of CFD calculations. In this light, CFD calculations focus on temperature distribution and fluid flow behavior. It becomes evident that a significant portion of the CFD results are concentrated on both temperature fields and fluid domains. Therefore, the classification of the 2D and 3D investigations is chosen and the positive outlines of 3D calculations are highlighted.

Primarily, 3D CFD calculations show the temperature gradients and the oil flow distributions more accurately than measurements. Chapter 2 also introduces the technical details and fundamentals of the power transformer and discusses the most crucial section in the active parts. Chapter 2 also presents different types of winding designs with special attention given to disc type winding designs.

The chapter also thoroughly examines insulation systems, encompassing both solid and fluid materials, to identify the key factors influencing insulation material aging. In addition, it discusses fundamental concepts of thermal modeling, including an explanation of the thermal reference diagram as outlined in the standard ICE 60076-2, along with the principal factors governing thermal modelling investigations.

The hot spot factor as a significant parameter in the concept of thermal modelling is introduced, along with a detailed explanation of the calculation method for this crucial parameter. The study also provides insight the primary cooling methods employed for the active components of power transformers.

Prior to evaluating the thermal performance of the winding, a crucial preliminary step involves identifying the heat sources and the distribution pattern of generated power losses. Hence, transformer power losses are discussed in chapter 2. It examines the classification of heat losses within transformers and shed light on the non-uniform heat loss distribution caused by radial leakage fluxes within the winding. In this light, different thermal stresses can be experienced during the operation of power transformers. Additionally, this work introduces various heat loss ratios, referred to heat loss eddy factors. The oil properties change by increasing the inlet oil temperature to account for the variations in the thermal characteristics of oil.

The impact of power loss distribution is explained, with a summary of the averaged winding temperature and the location of the hot spot temperature. Notably, the rate of the oil fluid flow directly influences the thermal behaviour of the conductors, particularly in the lower horizontal channels of the pass where the cooling oil share is relatively lower. Considering the last disc, the two central conductors experience higher heat losses. Despite the presence of higher heat losses at the top disc of the winding, non-uniform heat distribution results in a hot spot temperature located in the middle region of the pass (typically disc 3 or 4). This is attributed to a greater oil share in the last horizontal channel, which leads to improved convectional heat transfer between the oil and conductors. It is also concluded that the radial leakage fluxes increase at the ends of the winding, significantly influence on both the average temperature and the hot spot temperature in the winding model.

Chapter 3 focuses on the numerical CFD investigation, identifying the procedure of solving the governing equations within the discretized domain. The winding models are provided with the appropriate dimensions and designs. This chapter also indicates the grid sensitivity analysis on the designed grid structures. Besides, the details of creating mesh at the interfaces are introduced for disc type winding model and a distribution winding. This study employs the insulation materials to cover each conductor in model B, along with an in-depth exploration of the concept of employing an additional vertical cooling channel within the winding.

In section 3.3, the text provides the details of configuration of the CFD solver and the setup of the CFD software. It outlines the procedure for solving the energy and momentum equations, with particular attention given to the choice of convergence criteria. Especially in this study, the convergence criteria and root mean square (RMS) residual levels of 10^{-6} are selected during the solving iterations.

Chapter 4 illustrates the oil flow through the second horizontal channel redirects towards the first channel instead of flowing upwards as intended. Since the oil enters the first horizontal channel through the outlet area, the heat transfer process is impaired. The hydrodynamic arguments can explain the mechanism of creating local overheating and the reversed flow.

Although the widening the vertical channels facilitates the rise of the oil level and its filling of the vertical oil channel, it results in increased oil flow within the last three upper channels, leading to a more pronounced non-uniformly oil distribution. By evaluating the number of discs per pass, it is pointed out that fewer discs between two successive fluid guides aid in reducing hot spot and average oil temperature. This is due to a larger share of the oil being distributed among the fewer horizontal channels.

In Chapter 4, section 4.4 discusses the advantages of having additional vertical cooling channel in the disc type winding design. Based upon insights gained from previous investigations, it is observed that the middle section of the conductors tends to have higher temperatures. Therefore, the additional vertical cooling channels by the design of additional vertical cooling channel give different winding temperatures, although the winding size kept unchanged.

Comparatively, the winding model equipped with the additional vertical cooling channel demonstrates a hot spot temperature approximately 10 K lower than the design without it. Furthermore, the average temperature of the winding model equipped with additional vertical cooling channel is lower, and the temperatures are distributed with a lower gradient between the conductors. Due to the more uniform oil distribution, this design has no local overheating. Notably, the entrance of the bottom horizontal channels experienced fewer eddies that lead to the flow of the oil more smoothly through the horizontal cooling channels. In the winding design without an additional vertical cooling channel, the average temperature at uniform heat loss distribution is 8 K less than the non-uniform heat loss distribution.

Different oil Liquids are used to investigate the temperature distributions in Section 4.5. The performance of oil liquids is studied in the OD cooling mode at different inlet flow rates. It is noticed that at OD design cooling mode, the natural ester oil has a lower temperature distribution, and the average temperature of the conductors is lower than mineral oil. At inlet temperature of 80 °C, the negative effect of higher viscosity in natural ester oil is minimized, and the higher thermal conductivity provides a better cooling condition in the pass. The positions of the hot spot temperatures remain consistent when different oil liquids are used. Therefore, it shows that the position of the hot spot temperature is mainly related to the oil flow distribution, and the temperature can be changed by using different oil materials.

Chapter 5 examines the transient thermal behavior of a distribution transformer during a heat run test and the transient thermal behavior of a winding model in an oil container. This study used a three-phase 400 kVA distribution transformer to perform a heat run test. In this regard, the test methodology is discussed in Section 5.1.2.

Additionally, the disc type winding model A is used to determine the thermal behavior of the winding during experimental and numerical investigations. The CFD temperature distribution within the conductors is given at the starting temperature of 20 °C for the distribution transformer in model B. Although the HV winding conductors have the same level of applied heat losses, the hot spot temperature is located at the upper region, and the higher temperature distribution is observed at the last stack.

At t = 100 s, the temperature of the conductors exhibits a gradual increase, and the oil temperature changes slightly. In addition, oil velocity is not considerably altered in first-time intervals. At t = 1000 s, there is a noticeable increase in the average temperature of the conductors. The generated heat results in a decrease in the viscosity of the oil, leading to improved oil flow distribution and a considerably higher velocity within the channels.

In alignment of the evaluations in the temperature field, the hot spot factors are calculated for both consulted models. Initially, the hot spot factor in both designs increase at the first-time intervals and the oil reacted thermally to dissipate heat through convection. The oil circulation is then provided to address the thermal driving force which improves cooling performance and lowers hot spot factor.

Finally, the range of 1.3 to 1.7 is introduced as a reasonable estimation for the hot spot factor by ON-cooling mode. This estimation serves as a basis for assessing the overload capabilities of ON-cooled power transformers.

Future investigations can focus on the thermal performance of different oil liquids containing nanoparticles. The benefits of using different ranges of nanoparticles have generated interest to evaluate their performances. In this context, this work enriches the literature on CFD results in the winding and oil tank with the wide vary of boundary conditions. It also suggests the consideration of radiators for forthcoming studies. These future investigations can explore different radiator types with varying dimensions and numbers of external fins, examining them under natural and forced cooling modes.

Finally, the results of this study can be combined with future outcome results to develop a comprehensive GUI (Graphical User Interface) linked to the solvers to benefit both users and designer. The key influencing parameters can be given as an input into the software, allowing for enhanced thermal performance in various power transformer designs.

References

[1] CIGRé Working Group A2.38, Transformer thermal modelling. Technical Brochures 659, CIGRé Paris, June 2016. CIGRé Brochure.

[2] Fundamental of Power Transformers, Handbook Volume I1, Alstorm Grid 2015.

[3] X. M. Lopez-Fernandez, H. Bülent Ertan, J. Turowski, Transformers, Analysis, Design and Measurement, Taylor and Francis Group, Brocken Sound Parkway New York, Suite 300. 2013. ISBN 13-978-1-4665-0825-5.

[4] Y. Fradkin, Economics of transformer design, SPX transformer solution, November 2018.

[5] J. H. Harlow. Electric Power Transformer Engineering. The Electric Power Engineering-Handbook. Taylor and Francis, 3rd Edition, ISBN 9781439856291.

[6] B. Hochart. Power transformer handbook. Butterworths, London, 1st Edition, 1987. ISBN 0-408-02590-5.

[7] S.V. Kulkarni and S.A. Khaparde, Transformer engineering: design and practice, volume 25 of Power engineering. CRC Press, 2004. ISBN 9780824757281.

[8] Waukesha, Transformer core and transformer oil manufacturing, transformer coil winding, Online: https://www.waukeshatransformers.com/ power-transformers/manufacturing/core-and-coil/

[9] A. Weinläder, Thermohydraulische Untersuchung von ölgefüllten Leistungs-transformatoren, Hochschulschrift, University of Stuttgart, Göttingen, 2016.

[10] J. Zhang and X. Li., Oil cooling for disc type transformer windings - Part 2: Parametric studies of design parameters, IEEE Transactions on Power Delivery, 21(3):1326-1332, July 2006. ISSN 0885-8977.

[11] IEC 60554, Cellulosic papers for electrical purposes. Standard, International Electrotechnical Commission, 2001-11-08.

[12] IEC 60641, Specifications for pressboard and press paper for electrical purposes. Standard, International Electrotechnical Commission, 2008-07-21.

[13] IEC 60763, Laminated pressboard for electrical purposes. Standard, International Electrotechnical Commission, 2010-08-06.

[14] S. O. Oparanti, U. Mohan Rao, I. Fofana, Natural esters for green transformers: challenges and keys for improved serviceability, Energies 2023. Online: https://doi.org/10.3390/en16010061.

[15] W. Wu, Z.D. Wang, A. Revell, H. Lacovides, P. Jarman, CFD calibration for network modelling of transformer cooling oil flows – Part I heat transfer in oil ducts, IET Electric Power Applications, 2011.

[16] R. Küchler. Die Transformatoren. Springer Berlin Heidelberg, Berlin; Heidelberg, 2nd Edition, 1966. ISBN 9783642524967.

[17] Dejan Susa, Power Transformer Winding Losses, CIGRE Working Group, A2.38, Montreal, 2009.

[18] S. Tenbohlen, T. Stirl, M. Stach, Assessment of overload capacity of power transformers by online monitoring systems, IEEE power engineering society winter meeting, volume 1, 28.01.01, p.p. 329-334.

[19] IEC 60076-2, Power transformers - Part 2: Temperature rise for liquid-immersed transformers. Standard, International Electrotechnical Commission, 2011.

[20] V. M. Montsinger, Loading transformers by temperature, Transactions of the American Institute of Electrical Engineers, 49 (2):776-792, April 1930.

[21] R. M. Morais, W. A. Mannheimer, M. Carballeira, and J. C. Noualhaguet, Furfural analysis for assessing degradation of thermally upgraded papers in transformer insulation, IEEE Transactions on Dielectrics and Electrical Insulation, 6 (2):152-163, April 1999. ISBN 1070-9878.

[22] D. H. Shroff and A. W. Stannett, A review of paper aging in power transformers, IEE Proceedings C -Generation, Transmission and Distribution, 132 (6):312-319, November 1985. ISSN 0143-7046.

[23] IEC 60076-7. Power transformers - Part 6: Loading guide for oil-immersed power transformers, Standard, International Electrotechnical Commission, 2005.

[24] IEC 60354. Loading guide for oil-immersed power transformers, International Electrotechnical commission Standard, 1991.

[25] IEEE standard C57.91: IEEE guide for loading mineral oil-immersed transformers, American National Standard Std., 1995.

[26] F. P. Incropera, Fundamentals of heat and mass transfer, Wiley, Hoboken, NJ, 6th Edition, 2007. ISBN 978-0-471-45728-2.

[27] H. D. Baehr and S. Kabelac, Thermodynamik: Grundlagen und technische Anwendungen. Springer Vieweg, Berlin; Heidelberg, 15th Edition, 2012. ISBN 978-3-642-24160-4.

[28] Y. A. Cengel, M. Boles, M. Kanoglu, Thermodynamics: An Engineering Approach, 7th Edition, 2015. ISBN 978-935-316-5741.

[29] H. D. Baehr and K. Stephan, Wärme- und Stoffübertragung. Springer Vieweg, Berlin; Heidelberg, 8th Edition, 2013. ISBN 978-3-642-36557-7.

[30] VDI-Wärmeatlas. Springer Vieweg, Berlin, Heidelberg, 11th Edition, 2013. ISBN 978-364-21998-1-3.

[31] H. Vosen. Kühlung und Belastbarkeit von Transformatoren: Erläuterungen zu DIN VDE 0532. VDE- Schriftenreihe Normen verständlich. VDE Verlag, 1997. ISBN 978380072259.

[32] John D. Anderson Jr., McGraw- Hill, Inc., Computational Fluid Dynamics, The basics with application, 1995. ISBN 0-07-001685-2.

[33] B.R. Munson, T.H. Okiishi, W.W. Huebsch and A.P. Rothmayer, Fundamental of Fluid Mechanics, 7th Edition, 2013. ISBN 978-1-118-11613-5.

[34] H. K. Versteeg, W. Malalasekera, An introduction to computational fluid dynamics, 2th Edition, 1995-2007, Pearson Education Limited, England. ISBN 978-0-13-127498-3.

[35] A. L. Gerhart, J. I. Hochstein, Ph. M. Gerhart, Munson, Young and Okiishi's Fundamentals of Fluid Mechanics, 8th Edition. ISBN: 9781119317234.

[36] J. Ferziger and M. Peric. Numerische Strömungsmechanik. Springer, Berlin Heidelberg, 2008.

[37] D. C. Wilcox. Turbulence Modelling for CFD. DCW Industries, La Cañada, California, 1998.

[38] B. E. Launder and D. B. Spalding. Lectures in Mathematical Models of Turbulence. Academic Press, London, 1972.

[39] M. Heathcote. The J & P Transformer Book. Elsevier Science, 13th Edition, 2011. ISBN 9780080551784.

[40] G. Swift, T. Molinski, W. Lehn, A fundamental approach to transformer thermal modelling. I. Theory and equivalent circuit, Power Delivery, IEEE Transactions 16 (2) (2001).

[41] G. Swift, T. Molinski, R. Bray, R. Menzies, A fundamental approach to transformer thermal modelling. II. Field verification, IEEE Transactions on Power Delivery, 16 (2) (2001) 176e180.

[42] S. Ryder, A simple method for calculating winding temperature gradient in power transformers, IEEE Transactions on Power Delivery,17 (4) (2002) 977e982.

[43] J. Zhang, X. Li, Coolant flow distribution and pressure loss in ONAN transformer windings. Part I: theory and model development, IEEE Transactions on Power Delivery, 19 (1) (2004) 186e193.

[44] A. Oliver, Estimation of transformer winding temperatures and coolant flows using a general network method, IEE Proceedings, Generation, Transmission and Distribution, 127 (6) (1980) 395e405.

[45] J. Zhang, X. Li, Coolant flow distribution and pressure loss in ONAN transformer windings. Part ii: optimization of design parameters, IEEE Transactions on Power Delivery, 19 (1) (2004) 194e199.

[46] Z. Radakovic, M. Sorgic, Basics of detailed thermal-hydraulic model for thermal design of oil power transformers, IEEE Transactions on Power delivery, 25 (2) (2010) 790e802.

[47] A. Tanguy, J. P. Patelli, F. Devaux, J. P. Taisne, T. Ngnegueu, Thermal performance of power transformers: thermal calculations tools focused on new operating requirements, CIGRé Paris, 2004.

[48] P. Picher, F. Torriano, M. Chaaban, S. Gravel, C. Rajotte, B. Girard, Optimization of transformer overload using advanced thermal modelling, CIGRé Paris, 2010.

[49] H. M. R. Campelo, C. M. Fonte, R. C. Lopes, M. M. Dias, J. C. B. Lopes Network modelling applied to core Power transformers and validation with CFD simulations, CIGRé Paris, 2012.

[50] Nordman, H., O. Takala, Transformer load ability based on directly measured hot-spot temperature and loss and load current correction exponents, CIGRé Paris, 2010.

[51] O. Szpiro, P. H. G. Allen, C.W. Richards, Coolant distribution in disc type transformer winding horizontal ducts and its influence on coil temperatures, 7th International Heat Transfer Conference, 1982.

[52] K. U. Joshi and N. K. Deshmukh, Thermal analysis of oil cooled transformer, 21, rue d'Artois, F-75008 Paris, 2004.

[53] A. Weinläder, W. Wu, S. Tenbohlen, Z. Wang, Prediction of the oil flow distribution in oil-immersed transformer windings by network modelling and CFD, IET Electric Power Applications 6(2):82-90. February 2012.

[54] J. Zhang, X. Li, Coolant flow distribution and pressure loss in ONAN transformer windings-part I: Theory and model development, IEEE Transactions on Power Delivery,19 (1) (2004).

[55] J. Zhang, X. Li, Oil cooling for disk-type transformer windings-part I: theory and model development, IEEE Transactions on Power Delivery, 21 (3) (2006).

[56] J. Zhang, X. Li, Oil cooling for disk-type transformer windings-part II: parametric studies of design parameters, IEEE Transactions on Power Delivery,21 (3) (2006).

[57] J. Zhang, X. Li, Analysis for oil thermosyphon circulation and winding temperature in ON transformers, Power Engineering Society General Meeting, 2007. IEEE, 2007, pp. 1.

[58] E. Rahimpour, M. Barati, M. Schaefer, an investigation of parameters affecting the temperature rise in windings with zigzag cooling flow path, Applied Thermal Engineering, 27 (1112) (2007).

[59] E.I. Amoiralis, M.A. Tsili, A.G. Kladas, Transformer design and optimization: a literature survey, IEEE Transaction on Power Delivery, 24 (4) (Oct. 2009) 1999e2024.

[60] M.B. Eteiba, M.M.A. Aziz, J.H. Shazly, Heat conduction problems in SF6 gas cooled-insulated power transformers solved by the finite-element method, IEEE Transaction on Power Delivery, 23 (3) (Jul. 2008) 1457e1463.

[61] C.C. Hwang, P.H. Tang, Y.H. Jiang, Thermal analysis of high-frequency transformers using finite elements coupled with temperature rise method, IEEE Proceeding Electrical Power Application, 152 (4) (Jul. 2005) 832e836.

[62] J. Mufuta and E. Van, Modelling of the mixed convection in the windings of a disc-type power transformer, Applied Thermal Engineering, volume 20, pp. 417-437, 2000.

[63] K. M. Takami, H. Gholnejad, and J. Mahmoudi, Thermal and hot spot evaluations on oil immersed power transformers by FEMLAB and MATLAB software, Thermal, Mechanical and Multi-Physics Simulation Experiments in Microelectronics and Micro-Systems, 2007. EuroSime 2007. International Conference, pp. 1-6, 2007.

[64] C. Shih, Natural convection in electrical transformers, High Voltage Conference in China, 2001.

[65] E.J. Kranenborg, C.O. Olsson, B.R. Samuelsson, L.-Å. Lundin, R.M. Missing, Numerical study on mixed convection and thermal streaking in power transformer windings, Proceedings of the 5th European Thermal-Sciences Conference, The Netherlands (2008).

[66] A. Weinläder, S. Tenbohlen, Thermohydraulische Untersuchung von Transformator-wicklungen durch Messung und Simulation, Conference: Grenzflächen in elektrischen Isoliersystemen – 3rd. ETG-Fachtagung 16.09.2008 - 17.09.2008 in Würzburg.

[67] J. Smolka, CFD-based 3D optimization of the mutual coil conjuration for the effective cooling of an electrical transformer, Applied Thermal Engineering, 50 (1) (2013).

[68] F. Torriano, P. Picher, M. Chaaban, Numerical investigation of 3D flow and thermal affects in a disc-type transformer winding, Applied Thermal Engineering, 40 (0) (2012).

[69] F. Torriano, M. Chaaban, P. Picher, Numerical study of parameters affecting the temperature distribution in a disc-type transformer winding, Applied Thermal Engineering, 2010.

[70] A. Skillen, A. Revell, H. Iacovides, W. Wu, Numerical prediction of local hot spot phenomena in transformer windings, Applied Thermal Engineering, 36 (0) (2012).

[71] A. Gustafsson, Y. Jiao, Transformer winding oil flow rate and hot spot temperature, a straightforward relationship?, Proceeding, Corpus ID: 212668748, published in 2016.

[72] C. Rosas, N. Moraga, V. Bubnovich, R. Fischer, Improvement of the cooling process of oil-immersed electrical transformers using heat pipes, IEEE Transaction on Power Delivery, 20 (3) (July 2005) 1955e1961.

[73] C. Ortiz, A.W. Skorek, M. Lavoie, P. Bénard, Parallel CFD analysis of conjugate heat transfer in a dry-type transformer, IEEE Transaction in Industry Application, 45 (4) (Jul. 2009) 1530e1534.

[74] A. Lefèvre, L. Miègeville, J. Fouladgar, G. Olivier, 3D computation of transformers overheating under nonlinear loads, IEEE Transaction in Magnetics, 41 (5) (May 2005) 1564e1567.

[75] J. Smolka, A.J. Nowak, Experimental validation of the coupled fluid flow, heat transfer and electromagnetic numerical model of the medium-power dry-type electrical transformer, International Journal Thermal Science, 7 (2008) 1393e1410.

[76] J. Smolka, D.B. Ingham, L. Elliott, A.J. Nowak, enhanced numerical model of performance of an encapsulated three-phase transformer in laboratory environment, Apply Thermal Engineering. 27 (2007) 156e166.

[77] N. Schmidt, S. Tenbohlen, S. Chen, C. Breuer, Numerical and experimental investigation of the temperature distribution inside oil-cooled transformer windings, International Symposium on High Voltage Engineering, ISH 2013, South Korea.

[78] S. Tenbohlen, C. Breuer, F. Devaux, R. Lebreton, N. Schmidt, T. Stirl, Evaluation of the thermal performance of transformer windings by numerical investigations and measurements, CIGRé Paris, 2016.

[79] R. Hosseini, M. Nourolahi, G.B. Gharehpetian, Determination of OD-cooling system parameters based on thermal modelling of power transformer winding, Simulation Modelling Practice and Theory, 16 (2008) 585-596.

[80] X. Zhang, Z. Wang, Prediction of pressure drop and flow distribution in disc type transformer windings in an OD cooling mode, TPWRD.2016.2557490, IEEE Transactions on Power Delivery, 2015.

[81] M. Salama, D. Mansour, M. Daghrah, SM. Abdelkasoud, AA. Abbas, Thermal performance of transformers filled with environmentally friendly oil under various loading conditions, International Journal Electrical Power Energy System, 2020, 118:105743.

[82] M. Stebel, K. Kubiczek, G.R. Rodriguez, M. Palacz, L. Garelli, B. Melka, M. Haida, J. Bodys, A. J. Nowak, P. Lasek, M. Stepien, F. Pessolani, M. Amadei, D. Granata, M. Storti, J. Smolka, Thermal analysis of 8.5 MVA disc-type power transformer cooled by biodegradable ester oil working in ONAN mode by using advanced EMAG-CFD coupling, International Jounral of Electrical Power and Energy Systems, 136 (2022)107737.

[83] L. Garelli, G.R. Rodriguez, K. Kubiczek, P. Lasek, M. Stepien, J. Smolka, M. Storti, F. Pessolani, M. Amdei, Thermo-magnetic-fluid dynamics analysis of an ONAN distribution transformer cooled with mineral oil and biodegradable esters, Thermal Science and Engineering Progress, Volume 23, 100861.

[84] K. Dongjin, K. Kyosun, W. Jungwook, K. Yungsig, Hot spot temperature for 154 kV transformer filled with mineral oil and natural ester fluid, IEEE Transactions on Dielectrics and Electrical Insulation, volume 19, pp. 1013-1020, 2012.

[85] M Yamaguchi, T. Kumasaka, Y. Inui, and S. Ono, The flow rate in a self-cooled transformer, IEEE Transactions on Power Apparatus and Systems, volume PAS-100, pp. 956-963, 1981.

[86] P. H. G., O. Szpiro, and E. Campero, Thermal analysis of power transformer windings, Electrical Machines and Electromechanics, volume 6, pp. 1-11, 1981.

[87] M. Daghrah, Z.D. Wang, Q. Liu, D. Walker, Ch. Krause, G. Wilson, Experimental investigation of hot spot factor for assessing hot spot temperature in transformers, International Conference on Condition Monitoring and Diagnosis (CMD 2016), 2016.

[88] X. Zhang, M. Daghrah, Z. Wang, P. Jarman, M. Negro, Experimental verification of dimensional analysis results on flow distribution and pressure drop for disc type windings in OD cooling modes, IEEE Transactions on Power Delivery, volume: 33, Issue: 4, 2018.

[89] S. Tenbohlen, N. Schmidt, C. Breuer, S. Khandan, R. Lebreton, Investigation of thermal behavior of an oil directed cooled transformer winding, IEEE Transactions on Power Delivery, volume: 33, Issue: 3, June 2018.

[90] M. Pivrnec, P.H.G. Allen, K. Havlicek, Calculation of the forced directed oil-circulation rate through a transformer cooling system, IEEE Proceeding electric power application, Part C, 1340 (4) (1987) 306–312.

[91] O. Takala, Transformer load ability based on directly measured hot spot temperature and loss and load current correction exponents, CIGRé Paris, 2010.

[92] Z. Radakovic, K. Feser, A new method for the calculation of the hotspot temperature in power transformers with ONAN cooling, IEEE Transaction on Power Delivery, volume 18, No. 4, 1284-1292, 2003.

[93] T. Gradnik, D. Susa, and P. Picher, In-service accuracy evaluation of transformer loading guide models, 2012 CIGRÉ Canada Conference, Hilton Montréal Bonaventure, Montréal, Québec, September 24-26, 2012.

[94] Zhang et al. Oil-immersed transformer online hot spot temperature monitoring and accurate life lose calculation based on fiber Bragg grating sensor technology, CICED 2014, 2014.

[95] B. Feuchter and K. Feser, Online diagnostic of the thermal behavior of power transformers, International Symposium on High Voltage (ISH 1993), Tokyo, Japan, 1993.

[96] S. Tenbohlen, D. Uhde, J. Poittevin, H. Borsi, P. Werle, U. Sundermann and H. Matthes, Enhanced diagnosis of power transformers using On- and Off-line methods, results, examples and future trends, "CIGRé Paris, 2000.

[97] S. Tenbohlen, T. Stirl and M. Rösner, Benefit of sensors for on-line monitoring systems for power transformers, Mat Post, Lyon, France, 2003.

[98] G. Pudlo, S. Tenbohlen, M. Linders and G. Krost, Integration of power transformer monitoring and overload calculation into the power system control surface, IEEE/PES Transmission and Distribution Conference, Yokohama, 2002.

[99] Z. Radakovic, E. Cardillo and K. Feser, the influence of transformer loading to the aging of the oil–paper insulation, International Symposium on High Voltage, Rotterdam, 2003.

[100] Z. Radakovic, Numerical determination of characteristic temperatures in directly loaded power oil Transformer, ETEP, January 2003.

[101] Z. Radakovic, E. Cardillo and K. Feser, Algorithm of the microprocessor thermal protection of oil power transformers, IEE International Conference on Developments in Power System Protection, 2004.

[102] R. Vilaithong, S. Tenbohlen and T. Stirl, Improved top-oil temperature model for unsteady-State conditions of power transformers, ISH, Beijing, China, 2005.

[103] S. T. T. S. R. Vilaithong, Investigation of different top-oil temperature models for online monitoring system of power transformer, CMD, Changwong, Korea, 2006.

[104] W. A. Castillo, R. T. Jagaduri and P. K. Muralimanohar, Implementation of a transformer monitoring solution per IEEE C57.91-1995 using an automation controller, 65th Annual Conference for Protective Relay Engineers, 2012.

[105] S. Yun and I. S. Ch. Park, Development of overload evaluation system for distribution transformers using load monitoring data, International Journal of Electrical Power and Energy Systems, pp. 60-69, 2013.

[106] M. Djamali and S. Tenbohlen, Real-time evaluation of the dynamic loading capability of indoor distribution transformers, IEEE Transactions on Power Delivery, volume: 33, Issue: 3, June 2018.

[107] M. Djamali and S. Tenbohlen, Malfunction detection of the cooling systemin air-forced power transformers using online thermal monitoring, IEEE Transactions on Power Delivery, 32 (2):1058-1067, April 2017.

[108] S. Khandan, S. Tenbohlen, C. Breuer, R. Lebreton, Numerical investigation of fluid flow and temperature distribution on power transformer windings using CFD methods, ICCFD 2017, Berlin.

[109] N. Schmidt, Simulation and experimental validation of forced oil flow in power transformers, University of Stuttgart, Sierke Verlag, Göttingen, 2019.

[110] K. Karsai, D. Kerenyi, L. Kiss, large power transformer, Elsevier Science Pub. Co., Amsterdam, New York, 1987. ISBN: 9780444995117.

[111] F.R. Menter, Two equations eddy-viscosity turbulence models for engineering applications. AIAA Journal, 32(8): 1598-1605, August 1994.

Appendix

A.1. Considered Materials

Aluminum

$$\rho = 2702 \ \frac{kg}{m^3} \ , c_p = 903 \ \frac{J}{(kg.\,K)}, \lambda = 237 \ \frac{W}{m.\,K} \tag{A1.1}$$

Glass

$$\rho = 2500 \ \frac{kg}{m^3} \ , c_p = 840 \ \frac{J}{(kg.\,K)}, \lambda = 0.76 \ \frac{W}{m.\,K} \tag{A1.2}$$

Makrolon

$$\rho = 1200 \ \frac{kg}{m^3} \ , c_p = 1170 \ \frac{J}{(kg.\,K)}, \lambda = 0.2 \ \frac{W}{m.\,K} \tag{A1.3}$$

PVDF

$$\rho = 1800 \ \frac{kg}{m^3} \ , c_p = 1120 \ \frac{J}{(kg.\,K)}, \lambda = 0.2 \ \frac{W}{m.\,K} \tag{A1.4}$$

Viton

$$\rho = 1900 \ \frac{kg}{m^3} \ , c_p = 974 \ \frac{J}{(kg.\,K)}, \lambda = 0.18 \ \frac{W}{m.\,K} \tag{A1.5}$$

- **Mineral Oil (Temperature Dependent):**

$$\mu = \rho \times 10 \left(-4.726 - 0.0091 \times (T)\right), kg/(m.\,s), \tag{A1.6}$$

$$\rho = 875.6 - 0.63 \times (T), (kg \, / \, m^3), \tag{A1.7}$$

$$c_p = 1960 + 4.005 \times (T), J/(kg.\,K), \tag{A1.8}$$

$$k = 0.124 - 0.0000625 \times (T - 20°C), W \, / \, (m.\,k) \tag{A1.9}$$

- **Natural Ester Oil (Temperature Dependent):**

$$\mu = 0.134 \times 2.71^{-0.03\,T}, \text{kg} / (\text{m.s}), \tag{A1.10}$$

$$\rho = 935.93 - 0.6376 \times (T), (\text{kg} / \text{m}^3), \tag{A1.11}$$

$$c_p = 1715 + 6.75 \times (T), J/(\text{kg.K}), \tag{A1.12}$$

$$k = 0.1651 + 0.00009 \times (T), W/(\text{m.K}) \tag{A1.13}$$

A.2. SST $k - \omega$ Turbulence Model

Menter [111] introduced the Shear-Stress-Transport $k - \omega$ model which combines the $k - \omega$ model by Wilcox [37] and the $k - \varepsilon$ model by Launder and Spalding [38]. The connected development aims to unify the remarkable characteristics of the $k - \varepsilon$ model regarding open field flows with the preferential performance of the $k - \omega$ concerning flows close to walls.

k, ε and ω are modelling characteristics for turbulences with k equaling the turbulent kinetic energy, ε describing the turbulent dissipation and ω defined as the specific dissipation rate correlating to the ratio of ε to k. In the preceding section on turbulence modeling, the parameter k has already been established, which can be interpreted as the characteristic that describes the transport of a significant portion of turbulent kinetic energy through large eddies. Conversely, ε serves as a descriptor for the dissipation of turbulent kinetic energy, primarily occurring within smaller eddies. The turbulent viscosity μ_t can be determined according to the equations A2.1 and A2.2.

$$\mu_t = \rho \cdot \frac{k^2}{\varepsilon} \cdot F_{k-\varepsilon} \qquad \text{for } k - \varepsilon \text{ model} \tag{A2.1}$$

$$\mu_t = \rho \cdot \frac{k}{\omega} \cdot F_{k-\omega} \qquad \text{for } k - \omega \text{ model and the SST } k - \omega \text{ model} \tag{A2.2}$$

The definition of the introduced modelling functions $F_{k-\omega}$ and $F_{k-\varepsilon}$ are very different depending on the specific model. Typically, these models encompass constants, material properties and in the case of the SST $k-\omega$ model, velocity gradients are integrated.

To combine the $k-\omega$ and $k-\varepsilon$ models, Menter [111] introduced a $k-\omega$ formulation of the $k-\varepsilon$ model. Furthermore, a position-dependent function is created, which integrates the distance from any given point in the flow to the nearest wall. This function equals zero at the wall and one at far distance to the wall.

To complete the SST $k-\omega$ model, a weighted averaging of the modelling parameters is presented, taking into account the $k-\varepsilon$ and $k-\omega$ model with respect to the position dependent function. The equation below illustrates the fundamental concept for the transport equations concerning the turbulent characteristics k, ε and ω. The dominated parameters are given below. To address challenges in simulating turbulent flows, SST (Shear Stress Transport model) is employed as a combination of $k-\omega$ and $k-\varepsilon$ models.

As $k-\varepsilon$ is a high Reynolds number model, thus adjacent the wall region $k-\omega$ model is utilized, while in the region farther away from the walls, $k-\varepsilon$ model is employed. This approach helps to effectively manage turbulence simulations in various flow conditions.

Close to the wall, within the viscous sublayer, the $k-\omega$ model is applied. In the regions distant from the wall, the function is zero and only $k-\varepsilon$ model is employed, as outlined in [111]. The SST model requires the solution of two equations: one for the turbulence kinetic energy, k, and another for either turbulence dissipation rate, ε, or the specific dissipation rate, ω.

In a simple $k-\omega$ model, the modeling of dissipation energy is incorporated alongside the turbulent kinetic energy. The variables specify the turbulent Prandtl numbers for k, ε and ω, respectively.